William Henry Bragg
1862–1942

William Henry Bragg by William Rothenstein 1934

William Henry Bragg

1862—1942

MAN AND SCIENTIST

G. M. CAROE

CAMBRIDGE UNIVERSITY PRESS
CAMBRIDGE
LONDON NEW YORK NEW ROCHELLE
MELBOURNE SYDNEY

Published by the Press Syndicate of the University of Cambridge
The Pitt Building, Trumpington Street, Cambridge CB2 1RP
32 East 57th Street, New York, NY 10022, USA
296 Beaconsfield Parade, Middle Park, Melbourne 3206, Australia

First published 1978
Reprinted 1979

First printed in Great Britain by Western Printing Services Ltd, Bristol
Reprinted in Great Britain by Weatherby Woolnough,
Wellingborough, Northants

Library of Congress Cataloguing in Publication Data

Caroe, G M

William Henry Bragg, 1862–1942: man and scientist

Includes bibliographical references and index.

1. Bragg, William Henry, Sir, 1862–1942.
2. Physicists – England – biography.
QC16.B66C37 530'.092'4 [B] 77–84799
ISBN 0 521 21839 X

To the memory of my brother
William Lawrence Bragg
who planned this book with me

CONTENTS

ILLUSTRATIONS

ACKNOWLEDGEMENTS

I have needed help; and have had much given. I have many thanks to record.

Scientific advice has been my most urgent need; this has been freely given by Professor David Phillips of Oxford who has been writing a memoir of my brother William Lawrence Bragg. I have received further help from Professor Anthony North of Leeds where my father was Professor; he gave an admirable talk on my father at the University there in November 1975. I have received much help also from Sir Gordon Cox, who worked in the Davy Faraday Laboratory in my father's time and was until lately Treasurer of the Royal Institution where WHB spent twenty years as Director.

I could not have had better qualified advisers to guide me. I am profoundly grateful to them.

Next I must thank Sir George Porter, the present Director of the Royal Institution, for allowing me to spend days delving among the 'Bragg papers' in the RI archives, and thank the library staff who kept having to find things for me. Then I want to thank James Friday, who was archivist there during the early stages of my work, for his scientific and practical help and his enthusiasm which carried me along.

I have not yet offered my thanks to Captain Stephen Roskill RN who gave me sound practical advice from his experience as author, also specialised advice on the background to my father's work for the nation in two world wars; and I must also thank him for introducing me to the Cambridge University Press. I am grateful to Dr Richard Ziemacki of the CUP for his kind patience. The accuracy of Miss Edith Eales' typing has been a great help.

But still I have not mentioned my husband. He has pushed, dragged and encouraged me through the job; without his aid this book would never have come into being. I thank him most of all.

The following have kindly agreed to the inclusion of quotations:

The Editor of *Nature*: from an address by WHB at the John Cass Technical Institute in 1924 as reported in *Nature*.

The Deputy Secretary, University College London: from WHB's address at UCL in July 1935.

The Oxford University Press: from the Riddell Memorial Lecture entitled 'Science and Faith', delivered by WHB at Newcastle in 1941.

The Secretary, the Royal Institute of Chemistry: from the Gluckstein Memorial Lecture, 'Chemistry and the Body Politic' given by WHB in 1935.

The Keeper of the Public Record Office: to reproduce letters from among the Cabinet Papers CAB 21/829 and CAB 90/1. Crown Copyright is acknowledged.

The British Broadcasting Corporation: from a broadcast talk by WHB in 1938 entitled 'Moral Rearmament'.

Details of each source are given in the Notes.

I am indebted to the Curator of the Carlisle Museum and Art Gallery for consent to reproduce the drawing of WHB by Sir William Rothenstein.

GMC

Summer 1977

1

INTRODUCTION: 'WHAT SORT OF MAN IS A SCIENTIST?'

Science has put the trusteeship of the natural world
into the hands of man.

It is understandable to want to know:

(i) What sort of man is a scientist?

(ii) How does he come to make discoveries?
What is his urge?

(iii) How does he conceive of his responsibility
for his discoveries?

Perhaps some kind of answer to these questions can
only be found by making case-books of the lives
and thoughts of scientists, each man's life providing
his individual and partial answer?

One man's coherent set of answers is given in the
writing, work and thought of

WILLIAM HENRY BRAGG, KBE, OM, FRS

This was the 'foreword' written for the book which my brother William Lawrence Bragg and I planned to make together about our father. I composed it; he liked it; and I was pleased. Ever since we had written a short memoir of our father for the Royal Society in 1962, his centenary,[1] I had been urging my brother to write a 'Life'. I could not make him do it. He was 'too busy'; 'a son could not write about his father'. I would ask at intervals 'Why can't you?' until one day (and I remember the spot on the dull drive of a Corfu hotel – we had been out sketching together and were returning mellow and hungry) he agreed: 'We'll make a book together; but *you* must prepare the material.'

He wrote out lists of chapters. This was to be his part to relate, that mine. He would write of the years in Australia before I was born (seventeen years after him); I would tell of the later life after our mother died and I was my father's companion; and we would quote extracts from his writings.

For many months I searched through scrapbooks of press-cuttings and sorted old letters; read piles of reprints; and I waited. For my brother was always too busy, and still a little hesitant. At last, tired of waiting, I began to write, to show him what I had found. I read him my script; he got more and more interested. 'No lack of material!'; 'Yes, I remember. . .'; 'I hadn't realised'; 'I must re-write that bit': . . .and I knew him caught when one day he exclaimed, 'I wish I could get on with *our* book; it is the book I've always wanted to write!' He actually got the promise of a 'retired Professor's grant' for the typing of it. But he was struggling to finish his definitive history of X-ray analysis which Bells have published;[2] and in his 81st year his health was precarious. The week after he had finished writing that book he died.

My papers lay about gathering dust. At last I shook them and I shook myself. I was the one left with the closest personal memories. I would anyhow get them down. My father often used to talk to me about his research, holding me up as it were to see over a wall, and making me fancy for a moment that I understood; but I am no scientist. Perhaps someone would help me.

I knew exactly the kind of book my brother had wanted to make: it was to be a book not especially for scientists; for, as he said, the science work was already in the textbooks anyway. But it should try to show something of the mind and vision of the scientific worker, tell of our father's vaulting hope for what science might do to help mankind and his country, particularly during the crucial years of science's 'rise to power'; and relate the life-story of an unusual man as far as possible in his own words.

My father's life and career were unusual in at least three different ways. One day I was talking with Sir George Thomson about him; Sir George said the great question was 'why did he come to research so late?' Most good scientists, he pointed out, make their particular discoveries early, and then spend the rest of their lives exploring those discoveries more fully – unless life or a war jerks them into another channel where they make others. But my father did not start on research until he was over forty; and then in a year or two, within his subject, his name was known across the world. He gave his own explanation for the late start – he had 'never thought of doing research' – but this is astonishing in itself.

That is the first. The second most unusual point about his career is that he shared scientific discovery with his own son, a unique example of father and son sharing work which brought them a joint Nobel Prize.

And the third aspect is his public success: how did a man so retiring, so completely without personal ambition, become such a public figure in the nineteen-thirties?

To try and understand, one must follow his life from his very slow, humble beginnings, of which he wrote himself. As my father once said, speaking about Faraday and his discoveries: 'it is natural to think first of the man himself and of his purposes and ideals, for here we shall begin our understanding'.[3]

2

BACKGROUND AND EARLY LIFE UP TO 1886

The Braggs were rooted in West Cumberland; mostly they farmed, or sons went to sea in the merchant navy. A farm account book gives their daily dealings in the eighteenth century, parish records show marriages with Sibsons, Haytons, Gibsons, Irvings; and a miniature of a Jamaica planter with powdered hair in a green coat with brass buttons, an Elliott, shows one intermarriage with real gentry, and we still have some silver Elliott spoons.

But Braggs are really yeoman stock from that corner of England uncrossed by trunk roads, where there are few great houses, or 'week-end cottages', few gentlemen's residences; where the land is owned by the yeoman farmer. It is rich land and quiet, green meadows that breed good cattle, sloping from the Lake hills to the Solway Firth; and it was at a farmhouse called Stoneraise Place in the parish of Westward that my father, William Henry Bragg, was born in 1862.

He once took me there. Stoneraise is a substantial stone-built house with a slate roof, like most of the farmhouses in those parts. The back, with back-door opening straight on to the large farm-yard, is eighteenth-century; the front, with a panelled front-door, early nineteenth, and there is a lawn in the front with a fine large chestnut tree. I visited there with my father in the 'thirties, and have been there again lately. The feel of the countryside remains the same, of shrewd settled prosperity. In the course of each visit we went out to tea at a farm-house: were given plate-cake and rum butter in the 'thirties, rum butter and plate-cake and much beside in the 'seventies. In each case we admired old furniture treasured as heirlooms as we sat on items of a 'suite'. By now, indoors, the old fireplaces have gone (replaced by tiled surrounds), and the modern kitchen has come; in the yard new farm buildings rub shoulders with old barns and byres, looking flimsy beside them. The white clover my father and I delighted to smell has disappeared from re-sown pasture. But only this far has progress changed the look of the land.

My father took me to see the covered market at Wigton where his mother sold her butter and eggs. In an account of his early days he wrote:[1]

> I do not remember my mother very well, as she died when I was barely seven. Just a few scenes remain. I think she must have been a sweet and kind woman. I remember how one day I was sitting on the kitchen table, and she was rolling pastry, and how I suddenly found I could whistle: and how we stared at one another for a quiet moment amazed and proud of the new accomplishment. I remember something of a visit to the seaside at Allonby when Jimmy, four years younger than I, must have been just able to walk. For we were on the sands together, and being seized with the idea that bathing was the correct procedure at the seaside, I succeeded in undressing him and putting him into a shallow pool. Fortunately we were seen from the hotel window.
>
> I remember going home again, and being at the railway station (nearest to Allonby), and seating myself on the edge of the platform with my feet dangling over the rails: my mother saw me just as the train was coming in and rushed to pull me back. I remember the day before I first went to school, I was playing on the floor of the little parlour at Stoneraise Place, and my father, coming home from market, threw down a little brown paper parcel on the floor beside me. My mother exclaimed at this carelessness; it was a slate and slate pencils, and I think there was some breakage. There was a little holder for slate pencils when they got short, and I wondered what it was for. I must have been nearly five or just five. I could read moderately well, mother having taught me: I have a vague memory of that. Next day I was taken to the little school at Westward and as a test told to read the piece about 'George and his Pony'. Having got through that I was told to sit still, and that no more would be wanted from me for that day.
>
> I suppose I soon managed to find my own way to school, for I don't think I was often convoyed. Across the home meadow and up over the top field, then on to the main road; across that and down a side road which crossed the Wisa beck at the bottom of the hill and then mounted again to the top of the rise where the school-house stood, and the old

church and graveyard. From there one can see far over the country sloping down to the Solway, which is little more than a gleaming line at that distance, and on to the Scottish hills, Criffel standing out from them. I think the schoolmaster Hetherington was a really good teacher: I have heard so since. Anyhow, I took kindly to his lessons, and before I left for Market Harborough in 1869 when I was just seven I was fairly well on with the arithmetic. I was doing what we called 'practice' and the like. I suppose practice meant the compound additions, subtractions and multiplications of commercial practice.

Mrs Robert Bragg of Stoneraise, my father's mother, was born Mary Wood, the daughter of the Revd Robert Wood, vicar of the parish.

My grandfather's vicarage was of course close by: I remember one day when I determined to go and have dinner with them instead of going home. It may have been, probably was, a hankering after my grandmother's famous potted meat. But when I got home at night I was asked why I had not come home to eat pancakes: it was Shrove Tuesday! And pancakes were only cooked on Shrove Tuesday.

My grandfather was a fine old man, gentle and dignified. He became vicar about 1824 and died as vicar about 1884. He was greatly beloved. He was a great collector: and quite a county authority on its plants and birds.

He had a very old-fashioned little church. There was a three-decker pulpit; clerk at the bottom, my grandfather in a white surplice to read the prayers, which he changed for a black when he mounted to the top to give his sermon. The barrel-organ was in the gallery: my mother used to play it (by turning the handle) before she married. I heard someone say, when I was a boy, that the instrument was so crude that the tunes could only be played in a given order, but I cannot think it can have been quite so primitive. I remember the clerk marching down the aisle, disappearing to mount the gallery stairs, and reappearing to give out the psalm which he proceeded to sing (as a bass solo) while one hand turned the barrel and the other worked the bellows.

When he was young my grandfather not only took the service at Westward Church, but walked over in the after-

> noon each Sunday to take the service at Rosley [nearly] five miles away.
>
> I remember my grandmother Wood as a dear little cherry-cheeked old lady in a white cap. I have been told that as Ruth Hayton she had been one of the countryside beauties.

It was a small world my father was describing, seen through the eyes of a small child. He wrote these memoirs in 1927 and included a few details about his forebears.

> I know surprisingly little about my ancestors on my father's side. . . My great-grandfather Brown was in some sort of business in Workington or thereabouts. His wife died, and he married [again]. The children of the previous marriage then left him, and Lucy Brown, my grandmother, went to Belfast and on to Portaferry, a small town some little way south of Belfast. She married my grandfather Bragg. . . They had four children, William Bragg Bragg (Uncle William), Robert John Bragg (my father), James Brown Bragg (Uncle James) and Mary McCleary Bragg, who married [a Mr] Addison of Oulton near Wigton.
>
> When my Uncle William was twelve, he was the eldest, his father was lost at sea between Cumberland and Belfast and the family was left badly off. Uncle William was then at the Belfast Academy. . . He could not go on very far with his education, though whether he left school soon afterwards I do not know. I should think he went on a bit, because he knew his Latin fairly well. The family moved to Birkenhead at some time, because Uncle William was there apprenticed to a chemist, Uncle James went into an office and my father (some years later) went to sea. . .(See some old letters).

We have these packets of letters, neatly tied and annotated in WHB's hand. From them we find out a little more about Robert John's life as a merchant seaman.

There are his seaman's papers. The first is a properly cut indenture, dated 31 August 1846, apprenticing Robert, aged sixteen, to Joseph Bushby. The apprenticeship was for five years for a total of £30, £4 to be paid as wages in the first year, £5 for the second and so on. A Mariner's ticket (dated nine years later) describes him as 'fair, hazel eyes, 5′5½″ tall': he was probably less at sixteen when he shipped in the *Nereides* of Workington, 530 tons. In the *Nereides* he made a number of voyages to the East with mixed cargoes and did well. His indenture came to an end in the year of the Great Exhibition, and on the back is written his reference: 'Liverpool Oct. 1851. The within named Robert Bragg has served the term of this indenture with great credit, having conducted himself with sobriety and to the entire satisfaction of his masters & Joseph Bushby.' There is also, on a back corner:

'1847 An Act. of wages etc. advanced R. J. Bragg –

		£	s	d
May	To cash at Bombay R 12½ Rupees			
	at 2/2 exchange	1	7	1
Aug.Oct.	To cash in China D 5 dollars			
	at 4/10 exchange	1	4	2
	To – knife & southwester & sundries		5	0
Jany.	At Sea a pair of Boots		9	0'

The earliest of the letters from Robert was written when he had already served four years, and was just off again to Calcutta (he was twenty).

January 25th 1850

My dear Mother

I am very sorry to inform you that tomorrow morning 9 o'clock is the time appointed for our departure i.e. wind & weather permitting...

I called this afternoon at Bushby's office to get my boarding wages, there was only old Mr Bushby & the clerk in the office, and while the clerk was counting out my money, Mr Bushby took me into another room, & told me if I continued to conduct myself as I had done, he would make me 2nd mate next voyage but he would not tell me in what ship...

I have now got everything requisite to make me comfortable during the voyage, which I hope we will make in about 9 months.

Mr Bushby the other day took me into the cabin & asked me if all the reports were true which the steward had told concerning Captⁿ. Michael, such as, did I ever see him drunk, & such like, I told him that they were all false, he then asked me was he kind to us; I told him he was. The next day Captⁿ. Michael asked me what Mr Bushby was

> saying to me about him; I told him all, & he said 'Well Robin
> I knew that when you came back, it would be all cleared up
> & I told them so, I did think of leaving her, but however
> now since the truth has been told I will stay by her'.

The letter is written in a fine and clerkly hand and signed with flourishes. At the bottom is drawn a square and in it written:

> Mother Good Bye
> Mary Good Bye
> James Good Bye
> & I hope we shall soon meet again

with a tiny picture of a lady in a poke-bonnet shaking hands with a sailor, bundle on stick over his shoulder. William had come to see his brother off; William added on the back of the page:

> Dear Mother, I have had Robt.s company this afternoon. The
> last I suppose for some time... I hope and expect he will
> be a good boy, & obedient till we see him again which I
> trust will be in 8 months...

And writing again a few days later he says:

> It may be a brother's partiality but I believe, Robert is a lad
> that follows virtue for its own sake rather than the mere
> avoidance of vice.

This is the first evidence of the earnest interest and responsibility felt by William as protector of his mother and head of the family since the loss of his father. His firm hand and stern affection continued down the years.

The *Nereides* reached Calcutta and Robert wrote home:

> Calcutta, June 12, 1850
>
> My dear Mother, Brothers and Sister,
>
> You will no doubt be greatly surprised at not receiving a letter before this will reach you, but the fact of the matter is this, we have made a very long passage indeed, having been 130 days from Liverpool to Calcutta...
>
> We had indeed exceedingly bad weather for some time after we left and were nearly three weeks in and about the Channel, the whole of which time we had very heavy gales from the S. Westward, and upon the night you mention in your letter we had it very heavy indeed, so heavy that the Captain said if it continued much longer he would be

obliged to put into some port, as we were drifting right upon the land... [Robert's thought then turned to his home-coming:] But Mother do let *Polly* [his sister Mary] meet me in Liverpool, you cannot believe what pleasure even the thoughts of it give me, and it would be so pleasant for us all to go down home together (if it was so would not the people stare at us in church the next Sunday) now do Mother allow her, as I dare say we shall be at home some time in October...

Your ever affectionate Son and Brother

Robert John Bragg

Polly I shall expect to meet you in Liverpool and you Jamie at the station house. R J Bragg.

We do not know if Polly met him on his return from that voyage, but there is a letter of the following year which describes an amusing meeting.

Liverpool Wed. Oct. 22 1851

My dear Mother

I am happy to inform you of our safe arrival and that we are all well. We docked this morning in the Albert Dock. You may guess how surprised I was when I went into the office & was going up to shake hands with William, when Mr McGowan shoved me through the spring door (leading into the back office) into the presance [sic] of a young lady. I was just backing out again with beg your pardon, when she burst out a-laughing and then & not until then did I discover that it was Polly. She had been waiting at the office for me these two days...

It is now after 12 o'clock, & as I have had no breakfast I am going to get some dinner with Wm. but previously we shall have a warm Bath. I leave this in Mary's hands to finish off to you & am joined by Wm. in kind love.

Your Affete. son R. J. B.

This voyage ended his apprenticeship. He made one more trip in the *Nereides*, writing proudly to his mother just before he sailed:

Rockferry Nov. 12. 1851

My dear Mother,

I suppose William will have told you all the particulars of my proceedings today. I have received my certificate, signed articles as 2nd mate at £3 per month and got my luggage aboard the ship.

They reached Calcutta by March '52; 'I think the Captain seems well satisfied with his 2nd mate,' Robert wrote, and then, with Robert promoted to chief mate they sailed for home.

On 14 May the *Nereides* was wrecked in the Hoogli River. There is a touching letter from a maternal uncle to William dated from Workington 18 July.

> Dear Wm
>
> We have been since yesterday in a most serious state of Grief and anxiety but which today has been happily removed by the receipt of a newspaper stating that the chief mate and four of the crew of the *Nereides* have been saved, the account yesterday per Shipping Gazette was that only five men and pilot were saved. You will perceive by the accompanying newspapers that the Chief Mate is all right.
>
> Kind love...

After the letter had been folded a note was added on the outside 'I am so pleased I scarcely know what I have written'.[2]

There is no account by RJB of his experiences; but his certificate of discharge survives, endorsed 'Captain lost with ship', and there is an account of wages, paid up to and including the day of the wreck – but nothing after.

In 1854 he was sailing in the *Maharanee* – chief mate at £5 10 0 a month. There is a note to William, written 25 February from Dublin. It is about an investment: 'I signed the cheque blank as I did not know how much you might require.' William was looking after his brother's financial affairs. William looked after finances carefully all his life and did much good with them. His letters are often full of business and figures: crossed and spiky, hard to make out, not enjoyable like the flowing hand and gentle descriptions of little Robert.

WHB's own account contains only one paragraph about RJB's career at sea. It starts:

> My father retired from the sea in the late fifties when he would be about 25. He had done very well; and on one occasion he distinguished himself greatly. His ship was in St. John's Newfoundland, and the captain fell ill, or I seem to remember collapsed in drink. My father was first mate,

> took charge and brought the ship home, being then a very young man. [The ship was the *Peeping Tom*]... Money must have been left to the family about this time, because my father was able to retire from the sea: he bought a farm in Cumberland, Stoneraise Place, near Wigton, and settled down there.

Now comes the most important event in Robert John's life:

> He married Mary Wood, the daughter of the Vicar of his parish, Westward, in 1861... My father told me it took four rings to get my mother married to him. As he was walking up the road to the church he must have pulled the ring out of his pocket: anyway, there was no ring at the critical moment. One had to be borrowed from Uncle Robert Wood: and this my father and mother would not return to him, so a ring had to be bought for my uncle to replace it: finally there had to be a real new wedding ring. Years afterwards my father was walking up the road to the church when he picked up the original ring in the dust. I don't suppose road making was very thorough in those days in the country, and the story is possible.

Next year, Robert made an expedition to London to see the International Exhibition of 1862 with William and James. Mary was expecting a child quite soon and why he left her then we do not know, but probably William had arranged it all. Robert went first to Liverpool to pick up James, and wrote from there:

> My own dear Polly...[it is confusing that both his wife and sister were christened Mary and he called them both Polly. He told her how he and James had paid 1/- to go on board the *Great Eastern*; the letter ends]...
>
> Your ever affectionate husband Robert John Bragg.
>
> The first letter I ever wrote to you.

There followed a short note from Market Harborough where they joined William (short because he mistook the post time), and then the next letter was dated 26 June from London, with printed headings:

International Exhibition 1862
Railway Enquiry Office
& Letter Writing Room.

My dear Polly,

Your kind letter I duly received this morning and was very glad indeed to have such glowing accounts of how well you are getting on with your farming operations etc. etc. You will perceive I am writing from the exhibition, for which liberty I pay the sum of one penny, and have for that sum accommodation & pens and ink found me, for paper & envelope I pay two pence extra. We arrived here yesterday morning and have got lodgings at the Ryders Hotel, Salisbury Square, Strand and are very comfortable indeed. . . This is our second day, for we set off for the exhibition as soon as we had engaged our lodgings. Last night we stayed until they closed, what we may do to-night I cannot as yet say.

It is an immense place. To give an idea of it in a letter would be an impossibility, at least for me. Just imagine. We have been wandering about all yesterday and today, and we have not yet got through a quarter of it. And what a crush there is, yet still plenty of room. Yesterday there was upwards of 53,000 people here, and today I think is considerably more crowded. It took us all yesterday to go through the Agricultural Implement Department, and I am sure we have not seen half of it. I wish you were with me, you would have enjoyed it highly. . .

I do hope you will take every care of yourself and do please be lazy, that's just what I want you to be, at least for the present. Mother and William are very pleased indeed that I, or rather *we* have taken more land, and give me every encouragement. They are very fond of you, Polly and talk a great deal about you. But now don't be getting proud.

William is sitting behind me while I write, so I think I must conclude so as not to overtax his patience too much.

Excuse all errors, as there is such a din and racket, it is almost impossible to work correctly. With kind love to Mr and Mrs Wood [her parents] and Mary Ann, and last though not least to yourself my own darling.

I am,

Your ever affectionate husband

Robt. Jn. Bragg

William desires his kind love.

There is one more letter (dated 1 July) begging Polly to be careful. On the next day 2 July before he reached home, RJB's first son was born – my father William Henry Bragg.

A single letter of congratulations on his birth is the only record left in the family archives.

Two more sons were born to Mary after my father, Jack and James, and then, aged 36 years, she died. We have a coloured photograph of her, her visiting cards and the receipt for her wedding cake; three letters to Robert from Allonby (one of them giving her account of the bathing escapade) and one letter from Paris – she once went with relations for three weeks to Paris and wrote from the Hotel Meurice, rue de Rivoli, that, just arrived, they were 'a little tired but had just had an excellent tea'. And there is one more glimpse of her in a letter written to my father in 1931 by an old cousin, Maggie Wood. Cousin Maggie sent him an intricate silk patchwork quilt made by his mother and wrote telling how, as a small child, she had seen the pieces set out at Church Hill (the vicarage) 'on the best parlour floor – do you remember that room opposite the parlour?. . .Mary's mathematical head would do good service in the planning,' she wrote, and she described Mary: 'I see her as a gracious figure rather taller and more stately than Grandmother and so kind. . .she had a rich rather husky voice.' Mary Bragg died in 1869, but her 'mathematical head' lived on in her son, grandson, and great-grandson.[3]

While Robert farmed, William had been building up his chemist's business in Market Harborough. I visited the shop in 1971; it was much as he must have left it, run by the elderly daughter of the chemist who bought it a few years after William died: the old bottles were still on the shelves and in the small square mahogany drawers she had found some of Mr Bragg's labels. She had known people in the town who remembered Mr Bragg 'popping a man up on the counter and pulling his tooth out in no time'.

The shop was a corner one, and behind it was the grocer's shop in James's charge, opening on to a side street. William, James and their mother lived over the shop.

My father, whom I am calling WHB in these pages, wrote in his autobiography.

> My mother died when I was six or seven, I am not quite sure which year [he was seven and it was 1869]; in 1869 I was taken to live at Market Harborough. . .
>
> Uncle William had in 1869 succeeded in re-establishing the old grammar school in Market Harborough. It is a quaint structure raised on wooden pillars. The butter market used to be held underneath it. The newly appointed master, Wood by name, was an able man, I believe: and the school grew. I was one of the six boys with which it opened after a long interval. Perhaps it was because of my Uncle's connection with the school that at the end of the first year I was given a scholarship of £8 a year exempting me from fees. At the prize-giving – there were many more than six boys at that time, so that there was quite an assemblage – my name was called out and I went up to the desk to get the scholarship, not knowing what it was: I was puzzled and disappointed to go back empty-handed. The school was quite good and I got on quickly enough: in 1873 I went up for the Oxford Junior Locals and was the youngest boy in England to get through: I got a 3rd class, and was told that I would have done better but that the regulations forbade a higher class to anyone who did not pass in Church History; in that I failed, as also in Greek.
>
> We lived a very quiet life at Harborough. Before breakfast Uncle James and I went out riding for an hour to an hour and a half: we got to know well the villages round. Our longest rides would be to Kibworth or Kelmarsh, six miles away. I was not fond of riding for some reason, though I liked the morning air and I liked the pony. Ball games of all sorts have always interested me more than country sports: I enjoyed them very much, while hunting, fishing and the like did not attract me at all; moreover they never came my way. After the ride the day was filled with school and preparation for school: and an occasional walk. There were very few games in those days: as the school was a day-school, without

grounds. At the end of my six years there, we had a little football, which was a great delight. There were no parties for children: we never went to other people's houses, and no children came to ours. I think my Uncle was too 'particular' – he was indeed a refined and educated man – to let us fraternize with the children of the small shopkeepers, and as he was a shopkeeper himself, we were not asked to the houses of the lawyer, the parson and so on. I expect too that he did not feel justified in entertaining much, as he had to work his way from times of poverty, and it was not until later that he was in quite comfortable circumstances.

It was in 1876 that Uncle William built his own house, Catherwood House, on the Square, which it spoils. It is of smooth Midland red brick, gaunt and narrow with a carved stone frieze above the ground floor and gryphons protruding at the corners. It must have cost a lot; he was very proud of it.

He was very good to me, taking infinite pains with my lessons, especially the Latin: we hammered out the Aeneid with great difficulty in the shop, while occasionally he broke off to serve a customer. Our scanning was quaint. There are a few unfinished lines in the Aeneid, and we tried to save a few syllables from the line before so that we could get two hexameters, not realizing that the half line was incomplete. Mr Wood [the schoolmaster] was greatly amused.

Uncle William was a fine character. He was rather domineering and was not always popular even in his own family, still less in the town, which he tried to push along in the ways which seemed to him right and generally were so. He used to lecture us terribly, talking by the hour, and I suspect he was not to be shaken in his opinions by anyone. But he had great ideals and he always wanted to make us share them. He was unsparing of himself and of us in trying to nerve us to do our best: there was to be no slackness. For all that he was very kind, and he had lots of humour. When in later years he had mellowed, and I came down from Cambridge at intervals to spend vacations with him, the first evening was always uproarious because I had saved up all the jokes and stories for him, and he sat drinking it all in.

Uncle James was a dear kind man, very simple and

earnest, repressed and overpowered by Uncle William. His education must have been cut short very early, and he had very little knowledge to build on. But all his life he tried to improve himself, and his chief reading was Cassell's Popular Educator, in which he struggled to learn French and other things all by himself. His chief pleasure was the long ride on one day of each week when he went on 'Black Bess' to get orders at the country villages. The whole day was taken up with the journey, and he had dinner at one of the furthest points on the round. I remember him coming home on cold winter days very ready for the hot supper that was ready for him. On the morning of his round our ante-breakfast ride was omitted.

When I was taken in 1869 to live at Harborough there were my two uncles, my grandmother [Bragg] and Fanny Addison [Aunt Mary's daughter]. I don't quite know how my Uncle managed to collar both Fanny and myself. I suppose our respective parents were talked to and forced to give us up. Fanny and I were very good friends: much in one another's society of course, because as I have already said we were not on visiting terms with any other families. My grandmother was a real conservative, a Church person as against Chapel. [She had spent years in Ulster.] She was terribly put out one day on finding that the maid-of-all-work had taken us to a service at the 'Independent' chapel to which she belonged. Grandmother thought our souls were in some danger, so in the evening when she was brushing Fanny's hair and I was reading alongside, she went through the Apostles' Creed with us. – 'I believe in God the Father Almighty', she read out. Did we believe that? Fanny and I meekly said yes, not knowing how to say anything to the contrary, even if we wanted to. And then the anxious and rather stern old voice went on through item after item, and we always agreed. Did we believe in 'the forgiveness of sins'? We said yes. Did we believe in the 'Communion of Saints?' We said yes. And so on to the end. Then she closed the book, reassured. We must have been about eight years old at the time. Fanny was six months older than I was.

WHB seems to have been a very serious little boy, his natural goodness taking the form of trying to be good according to the standards of behaviour set before him, so different from the spontaneous behaviour encouraged nowadays. A letter to his father, written at the age of just seven, shows him as what some would now dub an unpleasing little prig. The letter is written in copybook hand.

Market Harborough. 1 Sept. 1869

My dear Papa

I am very pleased to write to you, and I hope you and brothers, Auntie and Willie, are quite well. I go out every morning with Uncle James; he has taught me to ride on the pony, and now goes beside me on the mare. When we return we have breakfast and Grandmama is very kind and always pleased with me. Uncle helps me with my exercises at night, so that I am in school before nine; the Master is very pleased with me, and I try to be a good boy and a good scholar. I like to learn Latin. Uncle says I count very well. . .

My dear Papa,

I am your dear son

William Henry Bragg

WHB's tale continues:

Well, after Grandmother died, Aunt Mary came and brought Willie Addison, Fanny's brother, to live with us. Aunt Mary had been keeping house for my father at Stoneraise Place. [She was the same sister Polly who had so gaily surprised her brother Robert in the Liverpool shipping office on his return from a voyage. She had had to leave her Addison husband; WHB thought it was because he drank.]

Aunt Mary was also overpowered by Uncle William. My happiest experience of her was when she took me to stay at Leicester during the week when I sat for the Oxford Local Exam in the spring or summer of 1873. We felt we were on the loose and were companions: I had not known her as such before. We stayed at Cook's Temperance Hotel, the first hotel of the great 'Cook'. My Uncle told me he remembered [Thomas Cook's] first excursion of a Sunday School party to Loughborough. . .[Aunt Mary and I] nearly always got on well together. She was sometimes rather a partisan over her son Willie Addison.

Willie Addison's arrival was not a happy event for me. In

fact my life became, as I remember it, miserable to a degree. We did not fit, and were never friends. . . I liked peace and was content to be alone with books or jobs of any sort, dreamy and lacking in enterprise outside the occupations I enjoyed.

In 1875 my father came to Harborough and demanded me: he wanted to send me to school at King William's College, I.O.M. [his brother-in-law had married the headmaster's daughter]. I think he became alarmed lest he should lose me altogether. Uncle William had entered me at Shrewsbury. [King William's] College had a pretty good reputation: somehow the boys won a fair number of scholarships; one or two or three each year at Oxford or Cambridge. As a school it was poorly found: the fees were very low, board and tuition were about £60 a year. I was at the School House under Scott. It was the plainest of the lot as far as living went: it was clean, however. [The printed prospectus for 1879 states 'each boy has a separate bed.'] Our meals consisted of breakfast and tea, at which meals we each had one piece of butter, as much bread as we wanted, and tea. No jam, milk or cake or bacon or eggs unless we provided luxuries for ourselves, or our parents paid extra, and very few did that. Dinner consisted of meat and pudding, both very poor: supper, a piece of bread and butter. Baths once a week: and one thing I look back on with interest, namely that we were locked out of our dormitories all day, so that we had no chance of a change when we played football, we just took our coats off and put on jerseys as far as I remember. But the place was a very healthy one, and after the first year or two, when the bullying was rather unpleasant, I was happy enough. I stood high in the school, and liked my work, especially the mathematics: and fortunately I was very fond of all the games and played them rather well. So, though I was a very quiet, almost unsocial boy, who did not mix well with the ordinary schoolboy, being indeed very young for the forms I was in, I got on well enough. As I grew older, I became more at home, and at the end I was quite popular I think, and as Head of the School I had some influence. John Kewley, [later] Archdeacon of Man, was my best friend. The masters were not very good: Hughes-Games the Headmaster was fairly able, not very popular: Jenkins, our mathematical master in my later years was a good fellow, keen and a good teacher; he did well for us. Curiously enough the strongest character was the French master Pleignier, old 'Plan', of whom the boys were more afraid than of any one else. He had ideas too; once he tried to teach us senior boys to write good English essays, but I fear I disappointed him.

The incidents that come back to me in memory are not likely to interest you much. Perhaps the most remarkable was the occasion about 1876 when the whole school was summoned for trespassing on the grounds of an old mill (Silverburn, I think) near Ballasalla. We all had to march down in crocodile form to the court house at Castletown and be tried by a Manx jury, consisting, so we boys believed, of Manx farmers. Of course they found us guilty. They could not put us all in the box, but they picked out a few of us, of whom I was one. I owned up to having been there, and when asked what damage I had done, I replied 'Nothing to speak of'. This was warmly greeted by the prosecuting counsel, who now thought he had got hold of a principal villain. He was disappointed when he found that I had only got inside, with others, the big waterwheel and made it go round by walking up it in treadmill fashion. You see, I really did like to be accurate, and my answer was strictly true. We were told we were all fined, and some of us, including me, more than the others, but I believe the fines were never collected.

During the first two years or so, we had long holidays twice a year only, the school periods being half-years. There was a break of a fortnight (I think) in the middle of the half-year, when it was thought to be not worth while for us to go home: and a number of us stayed on. We enjoyed that: we idled and played games and went picnics in which some of the masters' families would join in. Blackberrying was a common aim; we made blackberry squash by putting alternate layers of berries and sugar in jampots and squeezing them down, they were a substitute for jam.

When WHB was carried off to Market Harborough his brothers

Jack and James remained at home with their father at Stoneraise Place and attended school at Wigton. Later they both followed WHB to King William's College. Jack had had a serious illness as a small boy. WHB wrote of him:

> He was fairly strong during his school life at Wigton. He came to King William's College two years after me and got on splendidly. But he fell ill again, and was in hospital, i.e. the sick room in the School House, for some time. I was allowed to sleep with him there, as there was no nurse. He was often in great pain. I remember going down to the kitchen (really forbidden ground) in the middle of the night to heat plates to put on his stomach. Once the cook heard the noise of someone moving, and came down in déshabille and a state of great fright. The doctor thought he had an ulcer in his stomach: in these days he would no doubt have been operated on, but that was never done. He recovered, and went on with his work. I remember that to amuse Jack I got a pack of cards and asked two other boys to his room and we played whist. Unluckily the headmaster came in and was horrified. I believe our housemaster Trafford caught it badly for not looking after our morals more carefully! Cards were strictly forbidden.
>
> Jack got on well after I went to Cambridge, and came up to try for a scholarship at John's. He won an exhibition, a good one. But he never took it up. He fell ill again, with the same complaint, and the Uncle took him in at Market Harborough. He was very fond of him, so was every one of them in that house. He was anxiously looked after: I suppose that an operation was then looked on as a dreadful thing, and so no doctor advised it.

Jack died suddenly a few days before WHB sailed for Australia in 1885. My father always said he was the cleverest of the family.

> Jimmie came to King William College about four years after me. . . He showed a lot of mathematical ability and ended by winning an exhibition at Emmanuel. Jim is more like our father than Jack or myself; much more sociable. . . I have never been much with Jim: I left Stoneraise when he was three: I saw him during summer holidays at Stoneraise and

again when I used to go to the Isle of Man for holidays when I was at Cambridge. But we were apart most of the time. He is a most lovable person.

My Uncle James farmed for some years in New Zealand, and then spent a great part of his life working in a successful export business between Australia and England. He died in 1938.

WHB's account carries on:

In 1880 I went up to try for a scholarship at Trinity [Cambridge] and was awarded an Exhibition: this was in the spring and I was then 17. The authorities at Trinity thought, however, that I had better wait a year, so I went back to school. For that last year I was very much by myself in the work I did. I was head of the school, and one of the cricket eleven, so that apart from the work, I was well in with the school doings. But I did not do well in the work in that year, and when I went up again to try for a scholarship at Trinity – hoping to get something better than that which I had won before – I did not do so well as in 1880, and was told that I would have won nothing had I not been successful the year before. I think that it was bad for me to be so separate from the other boys in that last year: there was no competition and I was rather a difficulty to the masters because I needed special provision. But a much more effective cause for my stagnation was the wave of religious experience that swept over the upper classes of the school during that year. [The effect was emotionally severe, leaving a life-long imprint. His own description is quoted in chapter 11.]

I went up to Cambridge in 1881, taking the rather unusual course of beginning work there in the Long [vacation]. I suppose I was in Cambridge six weeks or so, July and part of August. But I forget the exact date. I had rooms in Master's Court. I appreciated thoroughly the beauty of the whole place; and I liked going to Routh's classes. It was lonely, because I was doing the unusual thing: and I had no companions. But it was good all the same. As a scholar of the College I went up every Long afterwards: it was always a jolly time. Very few restrictions: just the regular classes three times a week with Routh, and the preparation for them. After that tennis in plenty: boating on the river above

Cambridge, and the summer weather, and Cambridge looking its best. I tried during that preliminary Long to get through an exam that would excuse me the Littlego: and I failed in Latin, which seems to me now to be very odd, as I had studied Latin from the time I was seven, and given a lot of school time to it, and worked conscientiously too! I had to take the Littlego [the Cambridge Preliminary Examination] in November after all.

Cambridge gave me a good time of course: though I might have done much better if I had known more or been more easily sociable. I ought to have gone to lectures on other subjects than mathematics, and taken an interest in other things. It simply did not occur to me. I could not afford, or thought I could not afford, to join the Union or the Boating Club: which cut off a good many opportunities. I had none of those experiences of discussion of the world and its problems with other young men, which many men seem to look back upon with so much pleasure. I worked at the mathematics all the morning, from about 5–7 in the afternoon, and an hour or so every evening, and then bed fairly early. Every afternoon I played a game, generally tennis, or went for a walk: my tennis was fairly good, so that I always found people ready to play.

And he played hockey – he said they cut their sticks from the hedges. When Queen Victoria sent the Duke of Clarence to Cambridge she enquired for a quiet game for him. Hockey was suggested. My father carried a scar on his head caused by the Duke at hockey.

I changed my exhibition for a Major Scholarship in 1882, which gave me a standing in the College. I had the right then to join the Trinity Tennis Club without election, and to wear the strawberry and cream blazer, which was a source of pride. I sat in the scholars' seat in chapel, and took my turn in reading the lessons. At the end of the three years I took the Tripos: I was rather run down, and a little frightened, especially when I could not sleep the night before: a novel experience which shows that I was not really in a bad way. My Uncle got alarmed at my letters, and came to Cambridge to reassure me. I was afraid I had not done well in the exams: I remember the anxious mind as I walked up Senate House Passage to hear the results. When I heard my name called out as Third Wrangler, [i.e. third highest] I was really amazed: I had never expected anything so high, not even when I was in my most optimistic mood. I was fairly lifted up into a new world. I had a new confidence: I was extraordinarily happy. I can still feel the joy of it! Friends congratulated me: Whitehead [of *Principia Mathematica* fame] came and shook me by the hand saying 'May a 4th Wrangler congratulate a 3rd?' He had been 4th the year before. As for the Uncles!

During the autumn of 1884 I worked for Part III of the Tripos as it then was. I believe none of us did too well, but we nearly all got Firsts, because the Senior Wrangler did not do any better than we did, and they could not give him a Second. I was terribly proud because a publisher came and asked me my terms for solving all the problems in Smith's Conics, to go into a book of answers. I had other things to do, and had to say no, but I remember that in my mind I declined an offer which I thought might bring me in £5! Why, £150 would have been nearer the mark! It just shows how little I knew of matters outside my own line of work. I was in fact very much shut in on myself, unventuresome, shy and ignorant. And yet I enjoyed my life at Cambridge tremendously: I missed much no doubt, being the sort of young man that I was, but I gathered in a lot. University life is spacious and beautiful, Cambridge is a lovely place, and Trinity is something to be very proud to belong to. I loved it all, the work and the games, the place itself and the country round and all the incidents. In my last year or two I had a delightful set of rooms over the old Combination Room: the staircase was just opposite the entrance doors of Hall.

At the end of 1885 I was going, one morning, along the King's Parade to attend a lecture by J. J. Thomson at the Cavendish, and was joined on the way by the lecturer himself. I knew him pretty well at that time: he and Carey Wilberforce and I used to play tennis together. He asked me if Sheppard was going in for the Adelaide post. This was the professorship in mathematics and physics which Horace

Lamb was just resigning. He had been in Adelaide since the foundation of the young university [in 1874] and wanted to get back to England. I had seen the advertisement, and the magnificent offer of a salary of £800 a year. I said to J. J. that I had heard nothing of any such intention on Sheppard's part, (he was Senior Wrangler in my year). I was astonished at the question; it had not occurred to me that anyone so young might be eligible. Also the salary seemed too big for such untried people – I had a vague idea that £300 a year was more our style. Then I asked J. J. whether I might have any chance (as I was only 3rd to Sheppard's 1st on the list), and he said that he thought I might! So when the lecture was over I went and telegraphed an application – it was the last day of entry. A few days later I was summoned to an interview in London, to the office of the Agent General of South Australia. I found that I was one of three who had been sent for to be interviewed. One of the three – the name was Adair, I think, I had not heard of him – could not come, he was ill. The other, beside myself, was my late examiner in the Tripos, Graham! That was a queer situation, he and I sitting together in the waiting room.

The interviewers were Lamb, J. J. and the Agent General, Sir Arthur Blyth. They knew all about me, and the interview was short. I remember that they asked me if I regretted having applied, and I said with some astonishment 'Certainly not'. I think that if I had been more sophisticated I might perhaps have been less positive: I might have wanted to be amongst books and people in Cambridge: I might have wanted to work for a fellowship, though as a matter of fact my chances did not look well, because in 1883 the 2nd, 3rd, 4th and 5th Wranglers were all Trinity men, and in my year the 1st (Sheppard), 2nd (Workman), 3rd (myself) and 5th (Cassie) were all Trinity men; so I am glad that no sophistication prevented me.

I went back to Market Harborough, and that evening as Fanny and I were playing about on the piano, a telegram was brought to me. 'As new professor of mathematics and physics in Adelaide University, would I give some particulars of my career!' Well! You can imagine my delight! which was reasonable. An assured position, a salary beyond all expectation, a new country with all the adventure of going abroad to it, a break away from being a subject, to be now my own master. I took the telegram across to my Uncle at the shop: he read it, finished without a word the posting that he was doing, took me home across the square in the dark, and on the way he broke down. It had not occurred to me that the glorious success would mean to him a parting that he would feel so badly.[4] But I hope that his own pride in the result of what he had always worked for through me, carried him through. People used to stop and ask him if it was really true about his nephew, and he could answer and speak about his 'nephew the professor'! Perhaps too his excitement and pride were rather a strain on his feelings. I have still some of the letters he wrote to me when I left him and the others, and when I read them again I realize the strength of his interest in what was happening to me.

By the way, I forgot to say before this that the electors could have sent out a senior Wrangler of great ability, but he was not safe with the bottle. They thought however that they had better consult an Adelaide man who happened to be in London, and he was in favour of the young man who so far had kept off the drink. The Adelaide man was my future father-in-law.

WHB wrote in 1936 to congratulate J. J. Thomson on his eightieth birthday and reminded him of the meeting in King's Parade. J. J. replied on 17 December: 'I remember advising you to go in for the Professorship. I have always congratulated myself on having done such a good piece of work. I had forgotten the interview with Sir Charles Todd, but if that had anything to do with your marriage I builded still better than I thought.'

The next three weeks was a grand time! Preparations for the passage, new clothes, new outfit altogether: and there was a grant of £150 from the Agent General to cover it all. Visits of my aunt and myself to the outfitters in Cornhill (Silver & Co.?), visit to the shipping office with Sir Arthur Blyth, interviews, clearing up at Cambridge, farewells to friends and so on. I got a book or two on South Australia and read with eagerness about the place and its history. Then finally the Aunt and Uncle William came up to London the day before I sailed. I had been staying there for a short time; next day they saw me off at Tilbury and there I was away on the great adventure, thrilled by it. . . The boat was the *Rome*, then the largest boat in the P & O fleet, 4500 tons. It was a great adventure to me of course; it was a new life, and I was my own master, all by myself, and I enjoyed it to the full. There were nice people on board. One of them was C. H. Rendall,[5] who was particularly kind to me. I had a single berth cabin away up in the bows, and the Bay of Biscay lifted me up and down a dozen or twenty feet each pitch of the ship; so it felt anyway. I was ill and Rendall fetched me out and took me to a spare berth in his own cabin amidships. In a day or two I was all right and went back. Then on Monday morning (we had left on Friday) we were all a cheery crowd sitting on deck in the sunshine watching the coast of Spain go by. We saw Gibraltar from the sea. We landed at Malta and toured the island and dined on shore. We saw Pt. Said and all the clouds of pink flamingoes in Lake Menzaleb (I hope my facts are right!) We anchored in Lake Tinisak and were rowed across to spend an evening with passengers in the sister ship *Shannon*. We rode to Moses's Wells; drove round Colombo, and stayed the night at the Oriental and danced on the perfect floor. We had a fancy dress ball and so on.

The six weeks voyage came to an end and I was landed by tender at Glenelg.

My father's account ends soon after his arrival in Australia. It is rather strange how little there is about his father throughout his story. Ever since Uncle William had borne him off to Market Harborough, the Uncle dominated and organised his life. There was but the one occasion when Robert John faced William and demanded that his son should be sent to King William's College instead of Shrewsbury; it must have cost RJB a lot so to flout William's plans.

But WHB spent holidays with his father. He wrote 'Each summer we went to Cumberland, Fanny and I and Willy Addison after he joined us. That was a delightful change: my chief memories are the wild cherries and helping with the harvest.' Later there were holidays in the Isle of Man, for, eleven years after his wife's death, Robert Bragg gave up, sold Stoneraise Place, and retired to Ramsey, Isle of Man: he sold everything, from 4 Mahogany Beds (four post and half tester) to 'Washing Tubs, Dolly and Stand' as the auctioneer's advertisement reads, 'being a TOTAL SALE' and which offers 'a good opportunity to parties in want'.

So far as we know, RJB was content living in Ramsey; he occupied himself with boats again. When my father was made CBE in 1917 the *Ramsey Courier* recalled his father... 'Few men were more respected and esteemed amongst his fellows than Captain Bragg, and especially so with respect to the nautical community. He took a deep and practical interest in the maritime doings of the port and this can easily be imagined when it is stated that he was the owner of such clippers as the *Wizard* and the *Nerbudda*.' He died suddenly in 1885, aged 55 years.

My father's highest praise for anyone's personality was to say that 'he was a dear man'. I feel sure that Robert John Bragg, small and gentle, my grandfather, was 'a dear man'.

3

ADELAIDE: 1886–1909

Australia opened a new world for WHB. 'The going to Australia, to a new work and an assured position, the people I met there, the sunshine, the fruit and flowers, was a marvellous change for me. I know that I had been lucky enough in England, but I am not ungrateful when I say that going to Australia was like sunshine and fresh invigorating air.'

It might have gone to another man's head to be a professor at 24, a handsome bachelor with the large salary of £800 a year; but Adelaide's welcome only warmed WHB to sufficient confidence and hope that he might be able to do a good job there. He arrived early in 1886: he took up his responsibilities at the University and wrote a description to 'Uncle William'. You can feel the glow of pride in the Uncle's reply, 'I knew my boy would rise to his position'.

Adelaide University had been founded in 1874. Though vigorous, the University was far from being a highly developed organisation when WHB arrived. There was a story that once the students had waited in vain for their professor to deliver his lecture; he had gone off to the gold rush. And WHB used to tell how on one occasion he had helped to get out the classics marks when the examiner was incapacitated by drink.

WHB found the Physics Department was small – quite a good thing, as he had a lot of physics to learn. 'I tried to learn some physics on the way out,' he wrote in his own account. Elsewhere he recorded:

> I was Professor of Mathematics *and* Physics. I had never learnt any of the latter, nor worked at the Cavendish except for a couple of terms after I had taken my degree, it was supposed by the electors that I would probably pick up enough as I went along to perform my duties at the Adelaide University. So I read some Deschanel's Electricity and Magnetism[1]. . . In Adelaide it turned out that there was only one elementary class of respectable numbers and a

> little class of two students who were rather more advanced; and I managed to keep in front. Sometimes I had queer experiences. There was so much to do in the two subjects that I did not always have the time to rehearse before the lecture the experiments which the laboratory assistant had prepared. But my trust in the reliability of physical laws was always justified. On one humiliating occasion I asked the assistant if he could get me some sodium chloride for an experiment, only to discover a little later that the substance was common salt.[2]

It was not so strange then as it would have appeared later that a mathematician should be appointed to a Physics professorship. Many of the great nineteenth-century physicists had been trained on mathematics. Very little experimental physics was taught in the universities. Before the Cavendish Laboratory was built in 1870 experimental research had been carried on in any place available: Clerk Maxwell worked in an attic, Faraday in the old 'Servants Hall' in the basement of the Royal Institution: Lord Rayleigh had built a laboratory for himself at his country house. Until past the turn of the century research students at the Cavendish were still supposed to make their own apparatus. My brother has recorded in his autobiography his impression of the Cavendish in 1911:

> It was a sad place at that time. There were too many young researchers (about forty) attracted by its reputation, too few ideas for them to work on, too little money, and too little apparatus. We had to make practically everything for ourselves, and even at that the means were meagre. . .I had to manage with bits of cardboard and drawing pins, and a very poor [X-ray] tube worked by an induction coil.

So it is not surprising that when WHB came to Adelaide he found very little equipment in the laboratory there. He apprenticed himself to a firm of instrument makers in the town, learned to work a lathe and made the apparatus himself for his practical classes. This may well have been the origin of his intense interest in the design of apparatus for experimental research and his deep respect for tools. He only had two head mechanics during his working life, Rogers in Adelaide, and Jenkinson who joined him at Leeds and followed him to University College and the Royal

Institution. With each WHB had a bond of friendship built on the respect of one craftsman for another, and with each he designed apparatus elegant in its simplicity and fitness.

A series of letters to Professor Richard Threlfall at Sydney show WHB wrestling with problems in physics in his early Adelaide days. They illustrate one of his most striking characteristics; WHB could not rest until he had mastered some new idea completely, reduced it to a logical form which satisfied him, and expressed it in the simplest possible way.

> Monday Nov. 23 1891
>
> My dear Threlfall
>
> Thanks for correcting my mistake about the coefficients of self induction: I might have gone on bothering about that point for ever so long: for I should not have attacked Maxwell's calculations for some time. I have not read all Maxwell yet, I am sorry to say: I did not read any Electricity at Cambridge, you see, and I have not read systematically since; just read such portions as at any time I happened to want.
>
> December 3 ['91?]
>
> . . .I hope I am not boring you with my letters. My only excuse is that your replies help me a lot in getting to understand the electrical theories: and that I find them most interesting as well. You know such a devil of a lot more than I do.

WHB also had to learn to lecture. It is said that he was an unimpressive lecturer to start with, he would be too careful, too mathematical. But he must soon have acquired the art, for by the mid-nineties he was drawing such crowds to the University Extension Lectures that they had to be repeated. And while he was learning physics and how to lecture, he was also learning to enjoy himself.

Soon after his arrival WHB had set up house with a doctor, Alfred Lendon. Lendon had been a ship's doctor on an emigrant ship, proud to deliver on arrival the same number of emigrants he had started with, until it was remembered that ten extra had come on board on the way, or so the story ran. WHB describes how on the first day after he arrived:

> a friendly doctor Lendon called for me and drove me on his

> round in his victoria. We called at Dr Way's and I was refreshed with green figs, lovely I thought. We went to supper at the Observatory, and I met the Todd family for the first time. Such a jolly lot they were! Mrs Todd made the household of course. I was much impressed by her calm statement that she did not think she could go to the Government House party because she had not a dress fit for it. Such open and unconventional a confession was a surprise to me. I was marvellously fortunate in being thrown into a society of the Todds and people like them, so open and kind and good natured.

The Adelaide people were the first and second generation of emigrants, energetically making life in a new land: the old land was far away – but it was 'home'. A flag was run up on the Victorian gothic tower of the Post Office when the mails came in; English fashions were seized on; English ways followed. But the hierarchy was so different from what WHB had known in the English Midlands. The South Australian aristocracy were the station owners with hundreds of miles of land and rarely a sheep in sight, and the intellectuals were the University people: there were 'the poor' of course, and some very rough poor, but Adelaide society was urban, and dressy too, clothes being a creative art in a land where the other arts were not easily come by; and from old letters one gets the impression that they all spent a lot of time giving parties, going for elaborate picnics in waggonettes, later on bicycles, getting up something in aid of anything and calling round to see if their friends were all right in between the set 'At Home' days. Entertainment was home made. And at the centre of this warm hospitable kindly life were the Todds.

The Todds are very important to this story, because my mother Gwendoline was the fifth child in the family of two sons and four daughters.

Little Charles Todd (5ft 6in and rather plain, born at Islington, son of a grocer) had been sent out from Greenwich Observatory in 1855 to be Government Astronomer at Adelaide with the special job of installing the new electric telegraph system in South Australia. The voyage with his wife Alice aged 19 years, already pregnant, her maid and her piano, took three months. The town of Adelaide was only 18 years old.

First Todd established the telegraph line between Adelaide and Melbourne riding long journeys through the bush, often alone, to map the way. Then in 1872 he accomplished his dream, to link Adelaide to Port Darwin in the north by the 'Overland Line', and thence by submarine cable to Europe. Hundreds of miles of unknown desert had to be crossed by bullock waggon carrying poles and supplies. At last, the parties working from north and south met, and Todd was able to announce the completion of the line from a spot in the very centre of Australia, Central Mount Stuart, sending a message to his wife in Adelaide. Her name was given to their base, Alice Springs.

Astronomer, Inspector of Telegraphs, Postmaster General too – FRS by 1889 and Sir Charles KCMG by 1893, Todd was esteemed and loved in the colony; so used to being known that once, back in London, he jumped into a hansom cab and called 'Home'. There are many stories about him, of his strange second sight, of how the Adelaide postmen grew beards like him, and when he was an old man of how Granny Todd always tried to keep him at home for a day or two before she was giving a party, else he would invite anyone he met – or so my mother told us.

The Todds lived at the Observatory on West Terrace, an ample two-storeyed house with wide verandahs, surrounded by paddocks where the aborigines came to camp once a year and the men were given coloured blankets, and their lubras (women) white stockings, on the Queen's birthday.

Charles Todd was gentle with his family, Alice firm beneath her kindness. Christian duty in the station where God had placed you was her code, unselfishness the prime virtue. It was quite possible for a bunch of Todds all to go on an expedition together, none of them wanting to go, but each thinking the others did. But vitality and affection made it all fun and, quite apart from their position, their warm sympathy surrounded them with friends and gave them a happy confidence in life.

Before long WHB was a constant visitor at the Observatory; he was very popular with Lizzie, Charlie, Hedley, Maude, Gwen and Lorna, and their parents. The youngest daughter, Lorna, wrote many years later: 'Fierce arguments over religion and social subjects were the order of the day amongst the men. The irresponsible chatter of my sisters delighted [WHB] most. It was a revelation to a young man taught to weigh every word he uttered, and he blossomed under the cheerful and inconsequent atmosphere.' He was nicknamed 'The Fessor'.

Gwen was only 16 years old when WHB was first introduced at the Observatory in 1886, but in the next year she wrote him a letter. 'Dear Professor Bragg, Thank you so much for those lovely flowers you brought me, it is almost worth having the whooping cough to get them.'

In 1888 Gwen went with her brother Charlie to Tasmania for a holiday; WHB followed them and proposed to Gwen. She accepted subject to approval, and Charlie telegraphed her parents. The answer came back 'Say everything kind to both.'

Gwen and Will were married on 1 June 1889 and lived in love and loyalty together ever after. But the engagement time was not smoothly romantic. Gwen once told her daughter how she had not wanted to be engaged then at all, she was barely eighteen and life was fun and she did not yet want the responsibility of a husband and a home; but she knew she'd never find anyone else so good.

I have a bundle of letters from my mother to my father from this time. WHB, who never spent an unnecessary penny on himself ('it's good enough for me' he would say whenever we protested that some garment was old and ought to be replaced), loved giving presents. After they became engaged he gave Gwen and her sister Maude a holiday by the sea while he stayed behind in Adelaide to correct University exam papers and write 'an oration'. And then I think he got alarmed. Could he be managing to get on with his work quite well if he was properly in love? So were his emotions suitable to the occasion? He got dreadfully worried. His own letters have not survived, but there are Gwen's replies, and I dare to quote a little from them.

She comforted him maturely: 'You mustn't be always trying to love me very hard, you *will* find it very hard if you do. So just leave it all alone and if you feel you can't love sometimes as much as you would wish, just look at it philosophically and think it's nature. Leave it all to take care of itself. . .I cannot imagine what it is that makes you get in such a stew.'

The only worries of her own that she mentioned were that she could not spell and that she feared he would find her unmethodical. And all the while she was having a very 'jollie' time, bathing, sketching – and reading too. 'I have finished the Pickwick Papers and have begun Robert Elsmere [Mrs Humphrey Ward's novel which so impressed Mr Gladstone]. I sincerely wish I hadn't, that book is making me miserable, but I could no more leave off reading it now than I could fly; it's so horrid, Will, knowing what's coming, isn't it, and not being able to put it off for a single page. You feel so helpless about it, don't you.' Mother always felt things so.

But she must have managed to reassure WHB about his worries, for in another letter she wrote: 'It's awful nice to get your letters just before going to bed. Especially when they are such sweet cheerful letters as they are nowadays.'

Perhaps even loving is an art which requires a little practice. There is no record of WHB having known any girl before Adelaide days save his cousin Fanny with whom he was brought up; and once when Fanny visited Cambridge while he was an undergraduate he had asked permission from Uncle William to invite her to tea in his rooms. Uncle William replied 'NO'.

But the honeymoon was entirely enjoyable. 'How nice everyone is' wrote Gwen to a sister. They were so happy: 'it's something awful, we are always having to calm one another'.

They set up house in Lefevre Terrace with two maids who bossed GB. In those days you had to have maids, though WHB had not even a study to himself. He worked on the dining room table, uncomplainingly clearing away his papers when a meal had to be laid. William Lawrence was born in 1890, Robert Charles in 1891, but Gwendolen Mary (myself) not until 1907.

And here I want to explain about names. I have already referred to my father as 'WHB'. My brother William Lawrence has always been known to family and friends as 'Willie' or 'Bill'; he only used his second name of Lawrence after he was knighted, to distinguish himself from his father Sir William. I never called him, and cannot write of my brother as 'Lawrence', so I have decided to use initials. WHB is my father; WL my elder brother; and my mother GB. My younger brother RC (Bob) leaves the story early when he was killed at Gallipoli in 1915.

To get back to the University. In those early years WHB found the quality of his students very disappointing. He set himself to find out what sort of teaching they had been given at school, and began thinking how it could be improved. He gave an address on the subject at the University Commemoration in 1888 (this may well have been the oration he was preparing while Gwen was at the sea). *The South Australian Advertiser* for 20 December had a leader on it, several columns long, for what the Professor said was big news (although he was only twenty six); Adelaide was very proud of its University and its professors.

He was to present another long report on education ten years later in 1898, after a year's leave of absence in England with a commission to find out all he could about education 'at home' that would help with South Australia's problems. Again there were long columns in the press. These reports are too long to summarise here; there is more about them in Chapter 10. But it is important to note these first landmarks in WHB's lifelong crusade for education.

Having established himself at the University another man might have turned his thoughts to research. But I believe WHB was too modest, and perhaps a little too unadventurous. He wrote in 1927, looking back: 'Perhaps it may seem wrong that a professor of mathematics and physics should be so long content with the ordinary round of teaching and management, but there may be some explanation in the fact that I had had no laboratory training nor had I come in contact with research. I had indeed never studied physics as now understood'.[3] Anyway, the Universities did not require research from their Professors at that time. 'Do the next thing' was almost a motto for WHB: he saw teaching and some reform of education as his immediate job. He got on with his teaching.

But WHB followed keenly the new and exciting scientific developments in the Old World. He repeated experiments as far as he could with his limited equipment. He wrote his first paper in 1891 and sent a copy home to Sir Oliver Lodge. The paper was on 'The "elastic medium" method of treating electrostatic theorems'. Lodge wrote in reply:

5 March 1891

My dear Sir

I am delighted to be of any service to you in the way of seeing your MS through the press. I have sent it to the 'Electrician' as being of higher standing perhaps and altogether more suitable just now than the E[dinburgh]

> Review. I have not yet very carefully read it, but I see the kind of thing. It is just what I thought of once doing myself, and I am very glad to see it done properly.
>
> There is one difficulty in the elastic theory, . . . Hastily it looks as if you had not escaped this little quicksand, but the whole thing is an instructive attempt for students and analogies always break-down somewhere or other. . .

Sometime early in the eighteen-nineties, WHB started on his life-long career of popular lecturing. He gave University Extension lectures. A press cutting from 1895 reads: 'The deftness and success with which the Professor carried out his experiments and the exceedingly simple method of unravelling the mysteries of a difficult branch of science did him infinite credit, and proved the truth of the announcement to the public that the lectures would be of an elementary character, and the audience would not be supposed to have a knowledge of physics.'[4] These lectures were on Radiation, Heat, Light and Optics; one of the series was on 'Photography in Natural Colours' repeating experiments that Prof. Joly of Dublin had shown at the Royal Society in London, decades before colour photography became generally available. Adelaide was fascinated by the new ideas, captured by the lecturing charm of the Professor: his audience marvelled.

The lectures in the following year, 1896, must have taxed the audience more: they were on the newly discovered Röntgen rays. The latest reports had been received by cable 'on Tuesday': 'And that Adelaide people are keenly alive to the enormous importance of the most modern scientific triumph was shown on Wednesday night, when the University Hall was packed with an attentive audience, anxious and eager to hear the promised lecture on Röntgen rays by Professor Bragg. His Excellency the Governor was present, with Lady Victoria Buxton and suite; as also was his Honour the Chief Justice.' Much of the audience 'had positively to scramble for seats', many being turned away.[5]

WHB described the apparatus; a Crookes tube from England lent by a manufacturing chemist who had not the battery to use it; a battery belonging to the University, a coil from Sir Charles Todd's department. Next, 'he drew attention to a number of objects lying on the table'. . .Lady Victoria's spectacles in a box and a dead mouse in a matchbox, Mr Moir's hand lying on them, and then 'the rays were turned on for eight minutes'. And again, Professor Bragg 'placed his left hand over the plate'. The reproduction showed 'an injury the Professor sustained in years gone by to the tip of his little finger [he had all but cut it off in the turnip chopping machine on the Cumberland farm]... It is necessary to see the system worked to realize how wonderful it is'.

There was considerable doubt in many minds about the new ray's invasion of privacy; members of the clergy were protesting against 'the revolting indecency of the invention'; a London firm of outfitters advertised 'Ray proof underwear'. WHB set up an apparatus in the basement of the Physics laboratory and doctors sent their patients there to be X-rayed. Some risks must have been run. WL remembered clearly having a smashed elbow X-rayed when he was five, remembered the alarming crackle and the smell of 'ozone'.[6]

In the same year that Röntgen discovered X-rays, Marconi in Italy was developing the transmission of signals without wires, – only 23 years after Charles Todd had taken such trouble to lay those wires across Central Australia. WHB got very interested in wireless telegraphy; so did Charles Todd. Todd was a Fellow of the Royal Society, and WHB was to become one. They were scientific cronies. After Granny Todd died in 1898 the old man used to come and spend Sundays in the Bragg home and after lunch he and WHB often discussed the latest scientific ideas, – or just sat in companionable silence.

About 1899 they began to experiment in wireless telegraphy together. Lorna Todd, the youngest and unmarried daughter living at home to look after her father, told this story:

> I think I am right in saying that the first wireless pole to be erected in Australia was in the Observatory grounds. A receiving pole was put up on the sand-hills at Henley Beach. My brother-in-law [WHB] did much experimental work there. One afternoon I remember my father asked me to pack tea and drive with him to Henley Beach, saying he would send a wireless to say we were coming. I felt a very 'doubting Thomas'... However when we got within sight of the top pole on the sand hill there was my brother-in-law waving his arms and his cap – as thrilled as any schoolboy that the message had come through. It seemed a miracle.[7]

Of course WHB gave more Extension Lectures about wireless telegraphy. To quote from a newspaper cutting of 1899, 'He spoke very modestly of the experiments he is now carrying on with Sir Charles Todd'. Nevertheless, the reporter boasts, 'the experiments are being carefully noted in England'. The lecture was crowded. . . 'If any fault could be found with the Professor, that fault lay in the extreme simplicity of his language which made his subject so easy to understand that everyone carried home a clear impression of 'ether' waves.'[8] And I am sure that a proud Mrs Bragg explained to her friends those bits they had not quite understood after all.

They were pleasant years, hard work at the University, but there was time to play. WHB played tennis and won a Crown Derby cruet in a tournament, helped lay out a golf course and won a medal; a reporter wrote: 'Professor Bragg's golf is the result of an infinite capacity for taking pains, as during all his golfing career he has set himself to master individual shots by constant daily practice.'[9] He introduced lacrosse into South Australia and captained the N. Adelaide team. In the 'nineties bicycling became the rage.

Then there were the long summer holidays by the sea, weeks of hot sun and leisure – at Port Elliott where the breakers came crashing in, or on St Vincent Gulf where you could hear the whispers of a wave travelling along the beach from half a mile away, so WL described it to me. WL hunted for shells and made an excellent collection: he was always proud of having found a new cuttlefish and having it named after him, *Sepia Braggii*. He took to solitary pursuits, such as shell collecting, being rather dreamy, ahead of his age group at school and not good at games like his younger brother. Bob (RC) was not so clever, but was gay and popular, with his mother's intuitive knowledge of how to deal with people. Long years later, WL once said to me (on Euston Station platform) – 'You and I find *things* easier than people, Gwendy.'

And GB sketched on those long summer holidays. GB was very good, she had been perhaps the best student at the Design School. When she married, the master grumbled at a good artist wasted, since ever after painting had to be fitted between the demands of family and social life. WHB became fired with her enthusiasm and took up painting too. He addressed himself to the problem with the same humility he showed before any new venture: he practised Gwen's technique rather as he practised his golf strokes, and they sent their pictures to the South Australian Society of Arts annual exhibitions. In the *Advertiser* of 18 June 1896 the critic remarked: 'Among the successes of the Exhibition are the pictures, principally landscapes, in water color by Prof. W. H. Bragg and Mrs Gwen Bragg. . . Their mode of working and the results secured are so much akin that the name of either artist might be attached to any or all of them.' And the *Register* reported: ' "Afterglow" by W. H. Bragg is a peculiar and impressive little effort. . . There is a deal of patient work in W. H. Bragg's "A Little Wilderness"; it is a taking tangle of grass and growth, flowers and foliage.'[10]

Gwen sang too; sang Oratorios with enthusiasm and 'Gilbert & Sullivan' which was all the rage; WHB played the flute gently (to the end of his life I remember him playing old simple tunes towards evening when his work was done). And GB and her friends loved 'getting things up'. At a Garden Fête in the grounds of Government House 'The principal attraction. . .was the decorated parasol parade arranged by Mrs Bragg and Miss Todd. There were 50–60 competitors and these ladies, all dressed in snowy white garments and carrying parasols of infinite colour [decorated with flowers]. . .paced through the graceful figures of a Polonaise to the strains of Loti's string band.'[11]

The trip to England in 1897 has been mentioned. GB had never been out of Australia. Now she was to see the Old World: to be shown with her boys to the Uncles. WHB and GB were to visit Egypt and Italy on the way, while the boys followed in charge of GB's eldest sister.

The excitement was tremendous and preparations careful. Furnished with a set of carbon-copy books for keeping a diary and sending duplicate letters, they started the great adventure on 17 December 1897. The first entry reads: 'Dear old Granny at one end and Uncles at the other.' They wrote every day except when excitement was too great, and in the diary the handwritings alternate. At first they recorded anecdotes of passengers and weather, and then at 4.45 on a Tuesday morning Gwen heard the screw stop, leapt to the porthole and saw a line of twinkling lights along a shore 'and then gradually the light came, first a soft grey with the rows of palms sharply defined against the sky and then a blaze of rosy light'. It was Colombo. The colour, the jostling boats that came out, the din; 'Oh friends, I'll never forget it... I simply sat and gasped at the beauty of it all, it simply apalled [sic] me.'

Then on to Egypt. They stopped in Cairo at Shepheard's Hotel. GB enjoyed the famous terrace. 'I could have sat there for a week and not got tired, such a mixture as was always hurrying two [sic] and fro. Smart women with beautifully cut skirts on bicycles and attended by beautiful young men in knickerbockers, steering in between Egyptians (like Moses)... Turkish women riding straddle legs on donkeys...with tiny short stirrups so their knees meet well on top of the pommel...a Bedouin on a dromedary...an officer in a high dog cart preceded by sais who run in front to part the crowd...' And then the smart ladies staying at Shepheard's: 'They descend about 11 o'clock dressed for the day with generally a maid carrying furs, sunshade etc...you are simply amazed at the gorgeousness of the toilets [sic]...how their heads can hold the hats with all the plumes beats me.' WHB writes a few pages later: 'We hear since that Shepheards is now looked upon as very fast, and most nice people go elsewhere. Still it was great fun.'

They saw the sights, they shopped, they saw the Pyramids. 'This was the first time we had seen them and we felt a little odd' (WHB).

Then they boarded a steamer and went up the Nile. Gwen painted sunsets behind the banks in a tiny sketch book 4in × 2in. They visited tombs and temples on donkey back, learnt about dynasties and gods in Baedeker. And one evening, walking on shore in the dusk after their boat had tied up for the night, they reached the railway and stumbled over a pile of band instruments and cricket gear. The Warwickshires were going up to the Sudan war. The staff boarded their Nile boat; WHB and GB became very friendly with the General and his officers who were off to join Kitchener, and with the party of war correspondents. But progress was slowed by the bargeful of horses towed behind.

Leaving Egypt they crossed to Italy; ecstatic accounts reached Adelaide, and as soon as they had been read at the Observatory, Lorna Todd would jump on her bicycle to take them the round of friends in the town. The travellers bought photos of Pompeii, brown reproductions of Michelangelo and Raphael in Rome and Florence; met the boys in Marseilles and at last arrived in England.

Volume 4 of the Diary stops in the middle of a sentence on the last page, in Florence. Volume 5 has disappeared, so alas there is no record of the meeting with the Uncles, though WL kept a memory of the Market Harborough back garden, and of toiling all one morning to rob the stones from Uncle William's rockery to build a fort.[12]

The months in England were well filled; WHB visited schools, made exhaustive enquiry into English education for his promised report, attended a Teachers' Conference at Aberystwyth and with GB made a bicycle tour in Wales. Gwen and the boys stayed at Market Harborough with the Uncles, she shopped in London ('buy me anything nice that will go with my drawing room dear' wrote her sister-in-law from Adelaide); they visited Greenwich Observatory for her father's sake and were entertained by Sir William and Lady Huggins. But most important of all, WHB met and talked with fellow scientists, so that when he returned to Adelaide he felt closer in thought to the new work going on in England, for it was illustrated by memories of personal contact. The experiments on wireless telegraphy were the first fruit of the trip.

Soon after their return to Adelaide they built a house for themselves in East Terrace and called it Catherwood House, after Uncle William's house in Market Harborough. WHB designed it himself. GB could not understand plans, so he made her a model.

In the following years WHB went on performing his pleasurable duties, academic, social and public, and giving elegant popular lectures – a set on radium, another on the electron. Once he was sitting enrolling people for some course (he loved to tell this tale) and a well-known Adelaide matron came to put herself down. WHB couldn't remember her name. 'Your *full* name, please?' he asked; she bent down blushing, whispered 'Mary' and swept on. One year the Duke and Duchess of York (later George V and Queen Mary) visited Australia, and WHB had to help to show them over the University. The excitement over the royal visit was tremendous, and GB described it with ecstasy in an unused volume of the carbon copy series. The royal route was decorated; GB had decorated her brother's house on the route with giant paper bluebells reaching up to the first-floor windows. The Duchess inclined her head and smiled: 'Very pretty', she said.

By the turn of the century, WHB was 38 years old; he had been nearly fifteen years in Adelaide. He had followed, studied and explained new scientific developments, and repeated experiments; he had taken no active part in research. But in the beginning years of the new century his excitement was mounting. Many years later (writing after the death of Rutherford in 1937) he told of his feelings at that time: 'We felt. . .that we were watchers of a marvellous process in the alchemy of Nature, but we were watchers only: we seemed to be without power to control or modify the radioactivity which Rutherford had taught us to observe.'[13]

And then, it seemed almost as if fate gave him a push, and pushed him right into the ring. He was made President of Section A (Physics) of the Australasian Association for the Advancement of Science; he had to give an address; he had to think of something to say (so often his spur to new thinking). Years later, in 1926, he gave an account of this time. I quote it almost in full, for this was the introduction to his life of research. After describing the excitement of going to Australia he continued:

> For seventeen years I worked steadily in Adelaide. Then came another crisis. It had never entered my head that I should do any research work. I was to give the presidential address to Section A of the Australasian Association for the Advancement of Science; the meeting was in Dunedin, New Zealand in January 1904. I thought that I could make an interesting address if I spoke of the recently discovered electron and of the phenomenon of radioactivity.
>
> While reading up the subject, I came upon some results described by Mme Curie which seemed to me capable of only one interpretation, and that an interpretation which had not yet been suggested. It was known that when the radium atom broke up into two parts, one large and one small, the latter, which was really an atom of helium, was driven into the surrounding air, and these particles constituted what was called the 'alpha' radiation. Mme Curie described experiments which implied that all the alpha particles thus expelled went about the same distance.

This interested me greatly. All ordinary radiations fade away gradually with distance; the alpha particles seemed to behave like bullets fired into a block of wood. But, if this were so, the particle must travel in a straight line through the air, as the bullet does through the block. Now, some hundreds of thousands of air atoms would necessarily be met with on its journey. How did it get past? It could not push them to one side, for it was smaller than they, and the simplest experiment on the billiard table shows that a ball will not go straight through a crowd of others however hard it is hit. It could not dodge from side to side, continually recovering its original direction: that would require that the particle should be possessed of intelligence.

There was only one answer to the problem. The particle must go *through* the air atoms that it met. There must be a moment when two atoms, the alpha particle and the atom being crossed, *occupied the same space.* This was contrary to all the teachings that I knew. Still, it seemed to be right; and as a matter of fact, it *was* right. So I gave my address at Dunedin and explained this point in the course of it. I had never tested the hypothesis by any experiment, for I had no chance of making a test, I had no radium.

When I got back to Adelaide I was given the funds for the purchase of a small quantity by Mr Barr-Smith, a Mæcenas who often befriended scientific work, and so I was able to try my special experiment, and all went well. Indeed many other results came tumbling out, all of which fitted in with the new radioactive discoveries made by Rutherford in Canada. I found that helium atoms of four different 'ranges', as I called them, were shot out from the radium preparation, which must belong to the four different active substances that Rutherford had shown to exist.

Then I got a hint from Professor Soddy, who was passing through Adelaide, that I should dissolve the preparation in water, which would wash away three of the active substances but leave radium itself, the parent of them all. So I did, but, horror of horrors, as I brought my measuring apparatus up towards the radium in the way I had learned to do, there was *no radiation at all* when I was well within the old range. However, with a very downcast spirit, I

 1

The University
Adelaide S.A.
August 31. 1904

Dear Rutherford

I have lately obtained some curious experimental results in connection with the absorption of the α rays which are, I think, new.

I must first explain that, in pondering over the differences in the absorption phenomena of the α and β rays I have been gradually coming to the hypothesis that the whole matter may be explained by supposing that the α rays are not liable to deflection by collision as the β rays are. Both kinds of rays lose energy by spending it on ionisation, but in the case of the α rays this is the only cause of their being "absorbed", or since the latter is hardly a proper term to apply, of their being brought to rest: the rays are, of course, not waves of energy but particles. I know of course that this is rather contrary to your theories: and yet I think you will be pleased with what I want to tell you because my results are so beautifully explained on your theory of radioactive change, and supply a new illustration to it; although on the one point of the law of absorption they run counter to what you have said in your book.

Suppose then the α particles go straight ahead, never turning to right or left until their velocity falls to about 10^8: each α atom passing through other atoms like one meteor swarm through others. This is to my mind the

The start of a long correspondence: WHB to Rutherford

pushed the apparatus closer still, and closer; and suddenly a tremendous effect flashed out. The radium itself sent out the particles of the *shortest* of the four ranges, not the longest as I had thought; and, free from overlying impurities, was shooting with great effectiveness. My assistant Dr Kleeman, and I were excited enough!

So I wrote to Rutherford in Canada. It seemed a very long while to wait the necessary three months for an answer. I knew I had made an important discovery, and it seemed, surely, that someone must stumble on it before I could get in from the other side of the world. I was away in the country when the answer might be expected. The coach that brought the mails used to appear on the skyline of the hill at four in the afternoon, and for many days I went to the post office hoping for the reply. At last it came, and all was well. How pleased I was! And I have never forgotten that Rutherford took the trouble in the middle of all his own exciting discoveries, to write so promptly to someone unknown.

After that, research was part of my daily life of course.[14]

This was the letter that WHB sent to Rutherford in Canada:

The University, Adelaide, SA.
August 31 1904

Dear Rutherford,

I have lately obtained some curious experimental results in connection with the absorption of the α rays which are, I think, new.

I must first explain that, in pondering over the differences in the absorption phenomena of the α and β rays I have been gradually coming to the hypothesis that the whole matter may be explained by supposing that the α rays are not liable to deflection by collision as the β rays are. Both kinds of rays lose energy by spending it on ionisation, but in the case of the α rays this is the only cause of their being 'absorbed', or since the latter is hardly a proper term to apply, of their being brought to rest: the rays are, of course, not waves of energy but particles. I know of course that this is rather contrary to your theories: and yet I think you will be pleased with what I want to tell you because my results are so beautifully explained on your theory of radioactive change, and supply a new illustration to it; although on the one point of the law of absorption they run counter to what you have said in your book. . .

And this is Rutherford's answer:

McGill University
Oct 23 1904

My dear Bragge [sic]

I was very much interested in the account of your experiment which you forwarded me and congratulate you on the result you have obtained. I am all the more interested that I was independently attacking the question and had come to a very similar conclusion but had not your data. I hope you will publish your work in the Phil. Mag. as soon as you can for it mustn't be buried with the A.A.A.S. I think you have struck the right explanation of the absorption of α rays and that it is a mere matter of careful experiment along the lines you are following to disentangle it completely. . .

WHB had already sent his paper to the *Philosophical Magazine.* There is a letter to J. J. Thomson:

10 Aug 1904

My dear Professor Thomson,

I have lately come across some curious facts concerning the α rays, which as far as I know have not yet been published. . .[WHB describes them; the letter ends –] I sent last week a paper to the *Philosophical Magazine*, and hope to send a second in a little while. I am not quite sure that the Editor will have the faintest idea of who the sender may be: if you think they would be inclined to be suspicious would you put in a word for me?

Letter after letter followed to Rutherford, long letters describing his work, almost papers in themselves. In one, of 19 January 1905 (but misdated 1904) WHB enclosed a 34-page handwritten account of the work he had been doing on radiation, and the background of that work. He wrote:

It contains a few ideas and arguments which led up to my experiments, and might possibly be of use to you. To save your time I will make a little index and preface it to the paper, so that you can pick out anything you may want. I also send a few notes on various points, which I have not published anywhere; and which might be of service.

Note how anxious WHB was to share his new ideas with Rutherford; the conclusions are backed by ample experimental data. The 34 pages are written out in neat longhand, illustrated by careful diagrams. It was a beautiful job in every way. An answer from Rutherford dated 3 July 1905 starts:

My dear Bragg

I recd. your letter a few days ago. It is very good of you to send your results to me. They have worked out uncommonly well and have proved your theory to the hilt.

WHB was off, at long last, on his research career.

There were times of setback, of course. In another letter to Rutherford he wrote:

Adelaide
July 16 1905

My dear Rutherford

Thank you for your long and most interesting letter: it was good of you to spend so much time in writing to me when you have so little time to spare, and should really be giving all of it to your holiday.

You have indeed done a lot of work: I shall look forward to your description of it with the greatest interest. Your result about the velocity of the α rays when they cease to ionise is most surprising, as you say: I thought the critical velocity would be about 10^8 as for the β rays. And when I worked out roughly a law connecting ionisation with velocity I took this for granted. Of course this effort of mine falls to the ground, when the assumed hypothesis is taken away. It will be easy to make a fresh attempt with your new result as a beginning. Only as regards the paper I sent home, I am now a little sorry I spoke, as to this particular point! Perhaps the proofreader, whoever he may be, will point out that the work now needs recasting. It was only a side issue, anyway.

It was difficult being so far away:

> The University, Adelaide
> Dec 21 1905
>
> My Dear Rutherford
>
> I am very glad to have your letter of Nov 4: it has just come, the steamer must have been a remarkably slow boat! It is very jolly to have your comments on my paper: most reassuring and welcome. You see, I am a little out of the world here, and do not hear very much; and so I sometimes wonder whether those who understand the subject are approving what I have done. You and Soddy have been most kind, and have quite kept me going!
>
> Don't absolutely murder poor Becquerel! Our letters on the subject have crossed. I am delighted to find that they agree as to their theory; as to experiment, of course you alone can supply that. I won't discuss the point any more now: I shall look forward to seeing your reply. I wrote to Becquerel direct, in addition to writing to the Phys. Zeit. I was not sure of an address that would find him, but he is such a big man that my letter could hardly miss, I do hope your experiments come out all right.

The letter also shows WHB's respect for and dependence on Rutherford and Soddy, both of whom were younger than WHB.

In 1907 his papers appeared in *The Philosophical Magazine* and the *Philosophical Transactions of the Royal Society*. Then in the same year there appeared a headline in an Adelaide newspaper 'The new Australian FRS'. Rutherford had been one of WHB's sponsors.

It was a wonderful time, and the modest WHB was carried forward on a wave of enthusiasm. One letter to Rutherford dated 17 December 1907 is so excitedly confident as to seem almost uncharacteristic:

> I have a good deal of work to tell you about: I will take the most interesting first. I have got the decisive experimental proof that the γ rays are not pulses, but corpuscular. It is so simple that if you have not already found it out, you can go into your lab. and do it in half an hour if suitable apparatus happens to be lying about; and in half a day if the things have to be put together. In fact it has been sticking out a foot for the last two years! . . .

'Professor W. H. Bragg Fellow of the Royal Society'; there was now strong likelihood that he would be offered an appointment nearer home. In January 1907 Professor Soddy had written to Arthur Smithells, Professor of Chemistry at Leeds, to suggest WHB for the vacant Physics Chair.

> With Rutherford coming to Manchester. . .no doubt an effort will be made in Leeds to get a strong man also. Bragg . . .is I believe a thoroughly all round physicist and mathematician. But what has excited my intense admiration is the work on the α rays of radium. . . He has also aroused keen interest & enthusiasm in the University for the new theories, and by extension lectures throughout Australia. While there I was much struck with the spirit he had created around him.

But Rutherford coming to Manchester meant reorganisation at McGill. Rutherford suggested WHB for a new Chair of Theoretical Physics there and wrote about it to WHB. The two tentative offers, to McGill and Leeds, came about the same time. There is an undated draft of a letter from WHB to Professor Lamb saying:

> I have lately received a letter from Rutherford. . .[he goes on to describe the job at McGill]. I am writing to tell you this; for I have not been tied to secrecy in any way, and you have been so much my good genius that I feel I should like you to know.
>
> I seem to have got on to a certain line of research work which is worth pursuing further. It has been possible to go on with it during the last two or three years because I have made rather a big effort which I could hardly keep up: I have done it so to speak out of hours. During the last twelve months it has been easier because I have had no honours students. The applied science courses are too attractive [WHB's efforts had helped to make them so]. In the future therefore I am moving to a dilemma: if I have my full tale of advanced lectures, I have no time for anything but teaching: if there are no senior students I am not so much use to the University here and I have no helpers with the details of research. In any case I am rather cut off from others in the same line: intercourse with whom would help me very much, and might save me from many mistakes. It is a grand laboratory at McGill and there are some fine men there.

> They offer me at least 3500 dollars; and there is a right to a Carnegie pension after 25 years' service to which my tale here (22 years!) counts. I believe Montreal is a healthy place; and it is only 6 days from England where I hope that my boys will soon be at college. I am not too well up in all the particular subjects on which they want me to lecture: but I think I could manage them with a little preparation...
>
> I am writing to Rutherford this mail to say I will consider an offer with every intention of accepting.
>
> It would be a terrible wrench to move from here: there never was a kinder lot of people, nor a nicer little city: but perhaps it is good to be stirred up.

Sir Horace Lamb replied on 1 October 1907:

> ...It was very friendly of you to tell me about the McGill business... The conditions of the post seem very attractive, and the proposal is a great compliment, on which you are to be congratulated. I think you are quite right to consider it favourably. I should not have been surprised indeed if you had been tempted by the Leeds post, but the present opening seems decidedly more cheerful, and I daresay you would hardly care to go again through the uphill work of working up a place from small beginnings. *Per contra* I hear that Leeds are spending money on Chemistry, and it is possible that a new man might make demands for Physics with some success.

But a disastrous fire at McGill crippled the University's finance and the tentative offer to WHB could not be confirmed.

WHB accepted the invitation to Leeds.

So now life, work and interests in Adelaide had to be wound up. WHB had loved the life in Australia, and he had enjoyed success there, success in his University teaching, in popular lecturing, education work, in sport and even in art: he was a highly respected citizen, Governor of the Public Library, Museum and Art Gallery, a pillar of his parish church: he was on the Council of the University and School of Mines; and latterly success in research had come to him.

'I am very interested in this research work' a newspaper reports him as saying, 'and it is for this reason, and this only, that I propose to leave Adelaide'.[15] He says this baldly, but his thrill in the research shines out in his Presidential Address to the AAAS, given at Brisbane in January 1909 just before he left.

The title is 'The Lessons of Radioactivity', and this is a piece from it:

> It is clear that we are dealing [in radioactivity] with the most fundamental characteristics of the atoms, with the building material and not with the structure, with the inner nature of the atom and not its outside show; and it is this which differentiates radioactivity from the older sciences. You will remember how Jules Verne, in one of his bold flights of imagination, drives the submarine boat far down into the depths of the sea. The unrest of the surface, its winds and waves are soon left behind. . .[and the boat] reaches at last black depths where there is a vast and awful simplicity. Here, where no man 'hath come since the making of the world' the silent crew gaze upon the huge cliffs which are the foundations and buttresses of the continents above. It is with the same feeling of awe that we examine the fundamental facts and lessons of the new science.[16]

And Leeds was offering new opportunities for exploration and research.

Press cuttings report eulogistic farewells; 'Australia has sent home her Melba,' one paper declared, 'and is now despatching her Bragg'.[17] There were rounds of parties and tributes and presents.

Family farewells are unrecorded: it must have been specially hard, the parting with old Sir Charles. Perhaps the old man turned for interest to the seismograph that WHB had got him from England.

The family left in the *Waratah*, the new ship in the Lund fleet, in late January 1909 – Will and Gwen, Willie aged 18 years just graduated from the University, Bob aged 17 years and Gwendy (myself, just able to walk), the family nanny and a number of black dome-topped trunks. The furniture was to follow by sailing ship.

4

FIRST YEARS AT LEEDS: 1909–1912

We landed at Plymouth in March 1909. Letters to the family in Australia have not survived from that period; but for many years WHB kept in touch with Rogers, his Adelaide instrument maker. A packet of WHB's letters has come back to us. Rogers had been his working companion in the laboratory for twenty years, and on leaving Adelaide WHB wrote:

> I shall miss you dreadfully in my new laboratory. You have been such a wonderful help to me in all my work: it was not merely that you could do whatever I asked, but you could always do the little more and make a thing so delightfully fitted to its work that there was a perfect charm in setting it up to try it.

The next letter was from Plymouth, dated 8 April 1909:

> My wife and I went up to Leeds for a few days directly after we got here. We left our family in lodging at this place, Plymouth. The Leeds people are really very nice: all those I met anyway. The place itself is grimy, even the suburbs; but you can get out into beautiful country to the North. . . It was awfully cold for some weeks after we came here, snow all over England, but I am glad to say the children, baby and all, stood it well. . .I had a day with Rutherford in Manchester, and you can imagine it was very interesting! He showed me 250 mg of Ra[dium] in solution in a pump. They draw off the emanation as wanted. When the lights are down the tubes are all phosphorescent. He does not like Pye's cells: the leads are stuck through corks or something of the sort and they go wrong. One of his research students Geiger says he knows where to get good cells in Germany: and is writing for both Rutherford and me. [And he adds] The Leeds laboratories are very curious. . .I wish I might have you for a while.

The 'curious' laboratories were replaced in 1932, and WHB, opening the new buildings, spoke of how he had 'laboured for years in the temporary shedding which did duty for a physics lab and in which. . .work was often commenced on a winter morning with the thermometer at 40°F'.[1]

WHB wrote gently and as cheerfully as possible to Rogers. But GB was appalled by Leeds, by the dirt, by the smoky dark, by the rows of little poor back-to-back houses. The only bright spots on the scene were the whitened doorsteps, which one must step *over*, not *on*. She was horrified by the pasty-faced babies carried folded into their mothers' shawls, and the rickety children.

We lived in a furnished house for some weeks; my white toddler's clothes were changed twice a day, the boys had nothing to do and WHB was stricken by GB's despair. At the University (black Victorian gothic) things were no better. In Adelaide WHB had lectured on his own conception of physics – he was the only physicist of standing in South Australia. In Leeds he had to fit into a hierarchy, conform to a syllabus. He felt constricted. In Adelaide, students had drunk in his words; the Leeds students he failed to win – they stamped their feet; his lecturing went badly.

The happy, free life in Australia seemed so far away, so desirable; the friends, the sun, the position they had enjoyed. Perhaps only when he reached Leeds did WHB realise what a success story he had been living from the day of his arrival as 'the young professor', through busy useful years, to the adventure into research and fame which had brought him home – to Leeds. Adelaide had put WHB on a comfortable pedestal; now he was the colonial professor returned on the strength of his research; and the research ground to a halt. For three miserable years WHB clung to the corpuscular theory of X-rays he had formulated in Adelaide and fought a retreating fight with the wave theory of Professor Barkla. WHB was so seldom fierce, but misery sharpened his words and they argued in the pages of *Nature* until the Editor closed the correspondence, so my brother told me.

It was Rutherford's friendship during this time of the 'Bragg–Barkla Controversy' (as it has come to be known by historians of science) that helped WHB through the lonely years at Leeds. More than thirty letters passed between the two friends in the years 1910–1912 alone, and they met frequently.[2] One exchange shows WHB in an angry mood, rare with him, and revealing. Rutherford had written on 9 February 1911:

> I have looked into Crowther's scattering paper carefully and the more I examine it the more I marvel at the way he made it fit (or thought he made it fit) J.J.[Thomson]'s theory. As a matter of fact, I find I can explain the first part of his curve of scattering with thickness in *large* scattering alone. I believe it is only by the use of imagination and failure to grasp where the theory was inapplicable led him to give numbers showing such an apparent agreement.

WHB replied on 12 February:

> Your opinion of Crowther's paper agrees with mine: do you know, I think it is quite an immoral paper, very nearly dishonest, though I am sure the dishonesty was not intentional. One would think no paper could be more mischievous than one of this kind: but no doubt things right themselves in the end, and I may be a bit prejudiced, Apart from the fudge in it, I think Crowther is tackling the more difficult problem before the simpler...
>
> [So far he is trying to find some excuse, but then:]
>
> No: I think Crowther's paper is absolutely bad. I must say I have often found myself wrong when I have felt inclined to condemn utterly. But I think the censure is just, this time. I thought that paper by Allen in the Physical Review was bad, but then it was obviously so, and quickly brought correction. But Crowther's paper is a different thing. It has the worst faults an experimental paper can have, because it unctuously brings round a lot of facts to suit a theory backed by a great name, and it is so jolly cocksure: and the theory is only a half truth all the time...
>
> By the way, I see Barkla is a little puzzled to explain why his formula does not quite fit his recent experiments, & apparently he contradicts Crowther's recent Roy. Soc. paper.

But there were things to be glad about too:

> Manchester
> May 9th, 1911
>
> My dear Bragg,
>
> I am intending to go up to London on Wednesday afternoon to attend the Soirée, and hope to be able to attend

> Thursday's meeting. I hope that you will come down too. Sorry I can't see you in Manchester.
>
> I had heard something of Wilson's results, but did not know that he had photographed the trail. It is really a splendid piece of work, and will no doubt throw a great deal of light on some of the mechanism of ionisation. I am sure you are highly delighted at the way things are turning out in favour of your views. The X ray photograph is really remarkable. I am sure that the experiments are not easy, for I spent several months' work on the same subject in 1906 and for the same object when I was in Montreal; but my apparatus was so contaminated with radium that I could get nothing definite. It is really very fine to see the things one has seen in imagination visibly demonstrated.

In this miserable time a letter like this from Rutherford must have meant much to WHB, and the cloud-chamber experiments helped to support his views.

Professor A. C. T. North of Leeds University has written out for me his own view of the Bragg–Barkla controversy.

> In retrospect it can be seen that WHB's argument with Barkla concerning the nature of X-rays and γ-rays stemmed both from the design of the experiments that they carried out and from their own pre-conceptions. Barkla based his experiments on the suggestion of George Gabriel Stokes that X-rays were pulses of electro-magnetic radiation caused when the electrons in the X-ray tube hit the target. One property to be expected of wave-like pulses, as shown by James Clerk Maxwell's theories, was that X-rays scattered by matter should be polarized, just as are the light rays scattered by the air. Barkla's experiments showed that this was indeed the case. He also showed that elements gave out X-rays of a 'hardness' characteristic of their atomic weight, just as the visible spectra are characteristic of the elements that produce them.
>
> WHB's background was different. He was led to the study of γ-rays and X-rays from his other experiments on the range and behaviour of α-particles. In particular, he became interested in the 'secondary' β-rays produced when γ-rays passed through matter. He observed that these secondary

β-rays were shot out mainly in the forward direction; this one would not expect from an effect caused by a transversely-vibrating γ-ray. Moreover, the energy of the β-rays was characteristic of the hardness (penetrating power) of the γ-rays rather than of their intensity. The scattering of the γ-rays themselves was far greater in the forward direction than in the backward, whereas waves would be expected to be scattered equally in the forward and backward directions. All of these observations were apparently inconsistent with wave-like γ-rays, but they were just what one would expect from corpuscular particles impinging on matter. Unlike α-rays and β-rays, γ-rays were known to be uncharged and WHB put forward his hypothesis of a neutral doublet, 'an electron which has assumed a cloak of darkness in the form of sufficient positive electricity to neutralise its charge'.

Their properties of causing ionization as they passed through matter and other characteristics had shown quite clearly that X-rays and γ-rays were of a similar nature. Barkla and Bragg were agreed on that. It was in fact many years later, in 1922, that A. H. Compton showed that scattered X- (or γ-) rays consisted of two components, one scattered 'elastically' (without loss of energy) in a manner that preserved the characteristics of a wave motion; the other scattered 'inelastically' with lower energy and distributed in the manner expected for a particle. As it happened, WHB's experiments had been carried out mainly on γ-rays which had a high energy and for which the second effect predominated; and Barkla had worked with 'soft' X-rays for which the first was the more important. So WHB and Barkla had been led to their differing conclusions by observing different properties of the rays.

WHB's correspondence with Rutherford at the critical period of this controversy highlights many of the complexities of the debate. Rutherford's friendly humour was cheering. Here is one of his letters:

Manchester, Oct. 14th 1911

My dear Bragg,

The arrival of your papers this morning reminds me of the fact that I have not heard from you for an age [*viz.* the summer break]. I have, however, heard of your movements and of your visit to the British Association, and of the 'luminous and lucid' address you gave there; but I trust that the luminosity you excited was not responsible for the production of external radioactivity which I see Ramsay [President that year] brought before your notice at the meeting. I thought it was long since dead, and its resurrection seems to me quite unnecessary. I understand that you turned him down with great kindness but firmness.

As to ourselves, we spent a very pleasant holiday in Scotland, and I have been back for six weeks working hard, and have nearly got a research through on the heating effects of radioactive products. I am full of work, but otherwise reasonably contented...

When is Campbell going to publish his next paper on the δ-rays? His last one reminded me of stories which come out in the magazines in which the hero or heroine is left in a most thrilling situation at the end of the number. I thought the paper was very Campbellesque, but we must await with trepidation the next issue.

You will observe that the last Phil. Mag. was highly radioactive, to which end we contributed materially...

Yours ever E. Rutherford

There is a subtle pun in the last line – both Rutherford and WHB had a 'material' particle theory. WHB replied to this letter on the next day:

...It was awkward to have Ramsay making those remarks. There were wilder things which he said and were not reported: quite stupid, because he was talking for the sake of saying something I suppose. Lindemann says that Planck's theory now is that the atom (or one of its resonators) can take in energy of any wavelength at *any* rate but can only give it out in definite amounts. That is why I tried to make it clear how awkward it is to suppose atoms can store up energy till the right moment arrives and an electron is emitted with its full complement of energy characteristic of that wave length. Lindemann says 'relativity' is rather in the background at Berlin just now. Have you seen the papers on ultra-violet light by Millikan's people? They are in the Physical Review. They seem to upset entirely the conclusions of Ladenburg and of Hull on the relations between wave length and speed of electron. The whole thing is in the melting pot again and it is quite on the cards that the partial similarity between ultra-violet light and X-rays may be greatly interfered with. We shall see by later experiments.

On 8 December WHB wrote out for Rutherford's consideration a lengthy explanation of the 'neutral pair' theory which he had developed:

I must write to you some sort of description of that radiation idea; it is not altogether clear to me even now, but perhaps in writing it down it may improve...

As introduction may I put down that the main object of a radiation theory is to give a reasonable explanation of why there is such a thing as a definite 'black radiation', and then why the law of distribution is what it is. Then you have to explain the relation to temperature, and Kirchoff's law, etc., etc.

Some of these purposes are admirably fulfilled by the wave theory with its idea of resonance; but the theory is not successful at all points and there is some mechanism unrevealed. The 'one-one' theory of X-rays and γ-rays may reasonably be tried, not with the object of supplanting the other at all, but because it seems to supply some ideas which the other lacks, even if it is deficient where the other succeeds. And it is worth trying both and then looking afterwards for the perfect scheme. So let us go the whole hog with the one-one idea, on this understanding. Where there are electrons in motion there will necessarily be an acompaniment of 'quanta' balancing them... [He ended, ten pages later]

Would you mind just keeping these sheets together and letting me have them back sometime. I am not sure how much there is in it; in fact I can hardly believe Einstein has not got it all somewhere, only perhaps he may not have known all the facts of X- and γ-rays that we know now. If you want to talk about it to anybody please do; I don't want any secrets about it, if there are any secrets worth keeping! But do let me know what you think.

Rutherford replied:

Manchester, Dec. 20th 1911

My dear Bragg,

I received your letter a week ago but have been driving at full speed at my book, and so have had little time to devote to thinking of the account of your theory which you kindly forwarded me. I am afraid that I have not had time or energy to digest it thoroughly, but it appears to me to be an excellent idea which will repay going into carefully. It seems very likely that something of that sort is at the bottom of the whole business.

I went to the Cavendish dinner, which was a pleasant affair, and had a brief talk with Barkla. He seemed quite ready to believe that the energy of an X ray was concentrated; but does not like your material [is this Rutherford's typist's error for neutral?] doublet. He considers that an X ray must be a type of wave motion, and is to be regarded as the simplest form of light. J.J.[Thomson] also expressed the latter view. Barkla tells me that he has been put up for the R[oyal] S[ociety], so we will have to consider his claims next year...

Yours ever, E. Rutherford

P.S. I send you back as you wished your MSS on the subject. I was rather struck in Brussels by the fact that the continental people do not seem to be in the least interested in trying to form a physical idea of the basis of Planck's theory. They are quite contented to explain everything on a certain assumption, and do not worry their heads about the real cause of the thing. I must say that I think the English point of view is much more physical and much to be preferred.

Rutherford's postscript shows again how different was the thinking of British experimentalists from that of their more theoretical counterparts abroad, and indeed at home. WHB replied immediately to Rutherford on 21 December in a letter which so clearly expresses his ideas that it is best to print it in full:

My dear Rutherford,

Thanks for your letter and comments. I will try to make something of the idea: I think it has a meaning in it.

Of course the adoption of one model or another depends on what you are trying to picture. J. J. [Thomson] and Barkla object to the neutral pair because they want to connect X-rays with light and the neutral pair seems to abandon the connection. At one time (in Australia) I thought so too, and there seemed to me no help for it because the X-ray and γ-ray phenomena wore such a corpuscular aspect that the wave theory seemed to fail: and I did not see how light could be worked on anything but a wave theory. I had not heard of Einstein or I should have tried to discuss his theory at the same time. Or more probably I should have felt that I did not really understand Einstein and refrained from discussion altogether: and I am rather glad I did not refrain. But admitting all that, I believe still that the neutral pair is a useful idea, a makeshift it might be called but certainly no more so than the pulse theory, and less so in a direction of first importance.

The crux is that the X-ray and the β-ray behave like interchangeable forms: and the neutral pair is the simplest expression of that fact. The interchangeability is the central fact of the X and γ-ray work. The pulse theory does not directly suggest it or explain it or account for it. We may have faith that it will do some day: but it does not now, in spite of Sommerfeld's work and J.J.'s work & the rest. There is something big missing from the pulse model: and it is really that which makes the pulse theory almost abortive. It cannot express the central fact and so fumbles and gives no clear indications: experiments are as a rule far ahead of it. When J.J. in Cond[uctio]n of Elect[ricity] through Gases tries to deduce absorption and scattering from the electromagnetic equations, it *must* come to nothing because it uses expressions which naturally imply a spreading of energy in all directions from the shaken electron. It won't come right until the expressions naturally imply the conversion of the whole of one X-ray energy quantum into one β-ray quantum and vice versa. It is on that, that theory must concentrate. Meanwhile physical progress will be quickest if the interchangeability is assumed: and as I say the neutral pair conveys the idea so easily that it seems to me it is no question of anyone's liking it or disliking it exactly. They might almost say they did not like using some form of function which expressed a law satisfactorily because a possibly parallel law was best expressed by some other function. If the two laws are really partial forms of a greater law the right thing is to see how far each form of function can be made to cover the facts and then devise the complete form. That is how it seems to me: and I am quite prepared to abandon the neutral pair as soon as the better thing suggests itself: until it does, it would be foolish not to use it when it expresses a central idea as nothing else does.

Yours always, W. H. Bragg.

Rutherford continued the discussion in his reply of 23 December:

My dear Bragg,

. . .It was very good of you to write so fully about your views on the γ rays. I quite agree with your point of view and also that the really essential points of your theory are the corpuscular nature of the X ray, and β ray. I quite appreciate the utility of the physical conception in your doublet; but I would at any time be prepared if necessary to consider it a bundle of concentrated energy which has all the properties of your doublet. It is unfortunate that one has no evidence on the velocity of X rays, for it would settle the point definitely once and for all. I think you will find that I have treated your views in quite a handsome manner in my new work.

I must confess that Barkla occasionally irritates me by his blind insistence on the pulse theory. For example, he apparently regards the scattered radiation as new pulses, while I have the firmest belief, as no doubt you have, that they are the scattered primaries. I have an idea that Barkla is an excellent experimenter, but is not strongly endowed on the philosophical side.

I did not wish to give you the impression in my last letter that J.J. had made any remark on your doublet theory. As a matter of fact he did not speak of it but spoke of the connection of X rays and light. He is at present very much occupied over his canal rays, and as far as I can see thinks of little else. . .

Wishing you all a merry Christmas and a happy New Year.

Yours very sincerely,
E. Rutherford.

A few weeks later WHB summed up the position as he saw it with imaginative prescience. He wrote to Rutherford on 18 January:

> Supposing the identity of X-rays and light to be established, the supposition is this, I take it:
>
> The energy travels from point to point like a corpuscle: the disposition of the lines of travel is governed by wave theory. Seems pretty hard to explain: but that is surely how it stands at present.

The acrimony of the debate over the nature of X-rays was sad, yet in the determined pursuit of the 'opposed' theories much information about radiation was gained that might not otherwise have come out for many years. It was not until the 1920s and 1930s that the wave *v* particle dichotomy was finally resolved, through the re-examination of Einstein's neglected work on light, written in 1905. Yet in 1912 WHB had foreseen that some such resolution might be expected:

> . . .inconsistencies are difficulties of our own making. If one hypothesis links together a number of observed facts, and a second hypothesis a somewhat different number; and if we think the two are inconsistent, the fault must be ours. We must be stretching one or other hypothesis to breaking-point, and we must work in the hope of finding a new hypothesis of greater compass. Until we do so, we are right to use those which are more limited; it is the way of scientific advance. . .
>
> It is curious to reflect that Newton rejected the pulse theory for wrong reasons, and Huygens the corpuscular theory for reasons also mistaken. It is even more curious to consider how little their mistakes affected their work. Their theories were no more to these men than familiar and useful tools. Much of the heated argument in which we occasionally indulge arises from the failure to recognise that hypotheses are in the first instance made for personal use. We

> really have no justification for demanding that others should adopt the means which we find most convenient in the modelling of our own ideas.[3]

WHB stuck to his corpuscular theory until well after his own son had explained von Laue's X-ray crystal pictures in terms of waves, but when the new work on X-ray crystallography got under way, WHB ceased to write about the nature of X-rays. Both WHB and Barkla were right in their experiments. And as it turned out, both were also right in their interpretations. Later, WHB would often say: 'Physicists use the wave theory on Mondays, Wednesdays and Fridays, and the particle theory on Tuesdays, Thursdays and Saturdays.'

And his son wrote much later:

> The dividing line between the wave or particle nature of matter and radiation is the moment 'Now'. As this moment steadily advances through time it coagulates a wavy future into a particle past.[4]

But WHB could not be sure that his work would be vindicated in future years, and the confidence he had gained in Australia drained away. He felt he had come home on false pretences: he was wretched during those first three years at Leeds. GB was also miserable to start with, and WL, sent up to Cambridge for the Long Vacation term of 1909, was lonely and frustrated. Only RC at Oundle was happy; he was a gay and cheerful person who made friends readily.

There was one tragic incident which further darkened the first year at Leeds. We had travelled home in SS *Waratah* on the return trip of her maiden voyage. On the next return trip she was lost with all hands between Durban and Capetown. It was never found out what had happened. During our homeward passage WHB had been worried about the behaviour of the ship; he talked to the Captain: Captain Ilbery was worried too. She would travel in calm water with a slight list to one side, and then, without seeming reason, slowly roll over to the other. WHB went to the Lund headquarters in London when he got back and reported his anxiety. He gave evidence at the inquest after the disaster.

For GB things began to look up when she had a house of her own. Rosehurst was a pleasant square house with low pitched slate roof and a large garden, with carriage sweep and lawn. Red may, yellow laburnum and purple lilac leaned out of a shrubbery of smutty privet and laurel, edged with pink and white quartz stones which fascinated me. GB found two maids; Theresa played the concertina for me in the kitchen when Nanny was out, and Lizzie the cook told fortunes in tea cups. Lizzie also baked our bread. On two days a week the whole house was filled with the delicious smell of baking; no self-respecting Yorkshire household ate bought bread in those days.

And GB began to make friends. The silver salver in the hall piled up with visiting cards, and GB returned calls. Sometimes one of her new friends fetched GB in a dog cart and the ladies did an afternoon's calling together. Occasionally I was taken too, and waited with the groom while the ladies disappeared with their card-cases: sometimes they returned swiftly, the lady of the house was out and she was ticked off on their calling lists with satisfaction; there was time for another call to be fitted in.

Leeds society was rich. Successful manufacturers lived in over-filled houses with tiers of pictures in heavy gilt frames. They kept many servants and gave long dinner parties. Professor and Mrs Bragg dined with makers of railway engines, ready-made clothes, with brewers and steel-makers; shrewd public-spirited people. GB's warm-heartedness awoke response, and she found how nice so many were behind the expensive façade. She returned hospitality. She made a few mistakes: family tradition has it that, mixing up English game birds, she ordered one partridge for a dinner party.

Nor was all society just rich; for the second generation of success in industry often married 'county', and on the outskirts of Leeds were fine houses, which vigour and taste had filled with beautiful things, where old water-colours were beginning to replace Victorian sentiment on the walls, and windows looked out on herbaceous borders. The mistresses of these homes were proud of their gardens, and went out with baskets to snip the dead rose heads themselves. They also fostered culture. One kind lady, assuming a Professor's wife to be cultured, got GB elected to the 'Little Owls', a small select club of ladies who met – their successors still meet – to read papers in each other's drawing-rooms once a month. GB, of course, was quite *un*cultured, she always said she spent her youth avoiding being educated: but in the 'Little Owls' journal there is a record of a paper 'Famous Scientists of the Eighteenth Century by Gwendoline Bragg'; largely by WHB I should think, he always helped his family out.

She was also elected to the Art Club and there she had immediate success. Members went on sketching parties; they sat on stools under green-lined umbrellas and painted exactly what was in front of them. 'But it was like that' was the answer to any criticism of design. And she threw herself into social work. WHB wrote a description of Leeds life to Rogers.

Nov. 18 [1910]

> My wife is very busy always with many engagements: . . . there are all the University here to keep in touch with, and the people of the place. Also this is a surprising city for charity and good works, and everybody is expected to do something. When winter comes and there is no getting about the country they all set themselves down to work hard at all their jobs.

GB's special interest was the 'Babies' Welcome'; there were no state-run clinics at that time, it was an entirely voluntary organisation. If it is true that to do something for somebody is the best way to cure a dislike, then it was working for Leeds that finally cured GB's initial horror of the place – working for Leeds alongside the energetic friendly people who called with cards one day and rolled up their sleeves to cope with undernourished babies the next.

It is sad that WHB did not enjoy Leeds and its people more; GB cracked a nut which he failed to crack. For the first three years he was too miserable about his work, and the richness in Leeds made him uncomfortable, the poverty saddened him. He had golfing acquaintances, and a tremendous supporter in Professor Arthur Smithells; but socially he sheltered behind GB, a habit which continued till the end of her life. Some Leeds friendships made by GB have lasted into the second and third generation.

However, in our second year in Leeds, and to our great delight, we acquired a country cottage. I think it was to give GB an escape from the dirty city. Deerstones was in Wharfedale, above Bolton Abbey; a grey stone cottage which WHB rented for £10 a year from red-faced old Mr Johnson who drove a station trap. It was one of a group off the main road, under Beamsley Beacon. Below our cottage a brown beck gushed between water-worn boulders at the bottom of a beech wood, and then spread tinkling through meadows. A track came down from the moor to a ford and a single plank bridge, and the moorland farmer brought loads of bracken on a sledge down the track in autumn. He was called 'young Mr Moon' and he would knock on our door with his sheep-crook and say 'I'll be killing to-morrow, d'you want aught?'

I loved Deerstones passionately, and I saw the grown-ups happy there too. WHB enjoyed the simplicity and the quiet; it had a special peace filled with the rustle of beech trees and the sound of the beck; and GB baked her own bread and painted in watercolour. My brothers went sketching too, and friends came out from Leeds and I remember the living-room hung all round with sketches in brown paper mounts. GB painted purple heather and vivid green trees. In Australia when they were engaged, WHB had given her a new paint box, and in her letter of thanks she wrote 'That is a sweet little paint box, there is only one colour you don't use in it, the emerald green.' Now in England she found the emerald green most usable; she could not get over the green of England, delighting in it.

On 16 November 1911 WHB wrote to Rogers:

> I have been more busy writing than anything else lately, being under contract to do a book for Macmillan on the work I have done since those first days when we stuck that radium bromide in a lead block and got down a bit of the optical bench to raise it up and down. . .[5] And of course we have experiments going in the old ways. They have put me on the Council of the Royal Society too, so that will keep me going, because you are pretty well bound to attend the Council meetings and it is a day's journey to London and back.

But WHB was still gnawed by the feeling that he had betrayed Leeds' expectation of him. In this letter there is no excitement about work; he is writing a book about research he has done, not doing anything new. So when a letter arrived from Sir Richard Glazebrook, he considered another possible change.

Bushey House Dec 20 [1911]

Dear Bragg,

Would the Principalship of a new Colonial University (of British Columbia) have any attractions for you? My friend W. D. Caröe of 3 Great College Street, Westminster is advising the Government of British Columbia as to the buildings of a new University.[6] He has just returned with a commission to find them a Principal & asked me for some names. The man would have to organise the whole place, find Professors and establish them and advise the Government as to the steps they should take with regard to Educational matters. He must be sympathetic to Colonial ideas and tastes...

WHB wrote about the suggestion in a letter to Lorna Todd. Here I should explain a little about my Aunt Lorna. She was my mother's youngest sister who in the early Adelaide days would drop in on the newly married couple on her way back from school to get my father's help with her sums. She never married. In the expected Victorian fashion she looked after her father, old Sir Charles Todd, until he died. She had a long sad face, shrewd wits, an enchanting sense of humour, and an enormous gift for friendship. She did not need a career; people were her vocation, and she dealt with them with love, confidence and astringency. My father valued her judgement; hence his writing to tell her about the offer from British Columbia. His letter has disappeared, but Lorna's reply reflects WHB's feelings. She wrote on

Monday night... If getting enough research is beginning to wear on you and make you feel *responsible*, here is a chance to throw it up...for the sake of the concrete starting of a big thing...and if you are to go on gradually feeling more and more strained and disappointed in your work, it might be well to seize this opportunity.

But WHB was not despondent enough to be prepared almost to retire from the world of research. He asked Rutherford what he thought about it. Rutherford replied in a letter of 10 January 1913:

I think that if I were in the position that I tired of Physical work and had not an idea left to work on, I should consider it an admirable position to occupy one's declining years; but

> I quite agree with you that it would be very difficult to leave the Physical world at such an interesting time, when there is so much to do and so many interesting problems in sight.

Rutherford's advice only confirmed the decision which WHB had already taken to decline the suggestion of British Columbia. I think that WHB had only been making sure; it is fair to guess that in his heart of hearts he knew what he wanted. For by the end of 1912 the shut-in period at Leeds was over; he was out through a door which his son had opened for him. They were off on their great scientific adventure together.

5

FATHER AND SON: 1912–1914

WHB's Christmas letter to Rogers in Australia, written on 21 November 1912, told him:

> Billy [WL] is coaching and demonstrating at Cambridge, and has just brought off rather a fine bit of work in explaining the new X-ray and crystal experiments. He is to write an account of it in *Science Progress.*

This chapter deals with how this happened, and what grew from it.

To go back a little. While WHB was struggling at Leeds, WL was doing brilliantly at Cambridge (he already had an honours degree from Adelaide). He read mathematics as his father had done, and at the end of his first year won a major scholarship, doing the exam while in bed with pneumonia. Then, at his father's suggestion, he changed his subject to physics, which decision was to bear much fruit and some trouble. In WL's undergraduate days at Cambridge their mutual interest in science was very companionable. WHB wrote to Rutherford in December 1910:

> It would be awfully jolly if you would come over and see us, all of you. We have the boys home, but perhaps you are of the same mind as us that a bit of squash at this time of year is all the more fun. Could you come for the week end at the end of the year, the three of you? I would have said this next week end, but we are full up. I would have liked to have seen you before then to hear about the new atom, but I am content to wait if thereby we may see the family over here.
>
> The atom sounds very fine. My boy wants to know if he may hear about it too as he has been going to J.J.'s lectures and hearing all about the new doublet idea.

Pride in his son and enjoyment of all they shared in science must have cheered WHB through the doldrums of Leeds; they discussed new ideas and WL would write to his father about interesting lectures he had attended. Here is an early undated letter which records, incidentally, a memorable first meeting between WL and one of his close scientific friends:

> Dear Dad,
>
> I'm so glad you liked the notes on Jeans. I'll bring the notebook home and we can go over it. He has being doing the pr[essure] of radiation business lately, and according to him it is the most utter rot. I got an awful lot from a Dane who had seen me asking Jeans questions, and after the lecture came up to me and talked over the whole thing. He was awfully sound on it, and most interesting, his name was Böhr, or something that sounds like it.

The letter goes on to discuss what techniques they should use in attacking the new problems they were studying, and a further passage shows him following his father's way of thinking and foreshadows their later work together:

> If you manage to get the right answer $\frac{x}{e^x-1}$ form, it is because some bad mathematical brick has been dropped... Bohr pointed this out and said that was why J.J.[Thomson]'s treatment was wrong... One must either drop Maxwell's equations as holding down to very small bodies, or else *not assume in the mathematical treatment* that the app. final state reached inside a constant temperature enclosure was one in any way final, an equilibrium at all. The two things are inconsistent with experience which tells us that there is *not* an infinite amount of energy in an enclosure. (Emphasis is WL's.)

This letter foreshadows Bohr's later work as well.

WHB maintained his anti-mathematical bias for the rest of his life; and WL once remarked, 'You either have nasty physics and nice mathematics, or nice physics and nasty mathematics.'

WL supported his father in other ways. During the long controversy with Charles Barkla WL had not been entirely sympathetic with his father's point of view, the Cambridge people were not. Still, writing about mica diffraction in the early days of the X-ray work, he ended encouragingly 'I am sure Barkla has only got reflected spots, what an old muddler he is!'

WHB kept his son's letters carefully, and there is a goodly

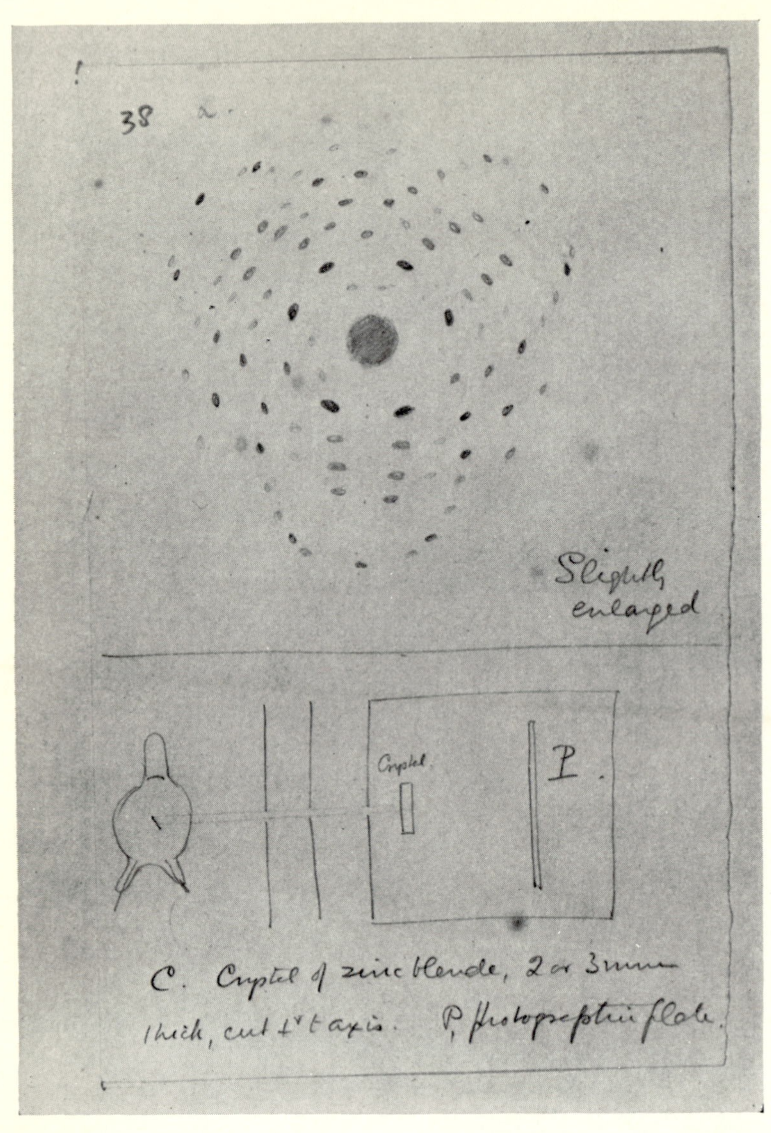

Facsimile of WHB's diagram of von Laue's 'curious X-ray effect' (1912)

collection, mostly from these early years; but WL tore up letters, there is only a handful left from WHB to WL.

In 1912 WL took his degree, a First; he was to stay on to work in the Cavendish Laboratory.

The debate on the nature of X-rays was still dragging on; but in 1912 von Laue in Germany published a paper with Friedrich and Knipping, which, coming a good decade before physicists saw that waves and particles were not mutually exclusive alternatives, was soon accepted as decisive evidence for the wave nature of X-rays.

Sir Gordon Cox has described to me the interesting way in which this paper originated. P. P. Ewald had gone to von Laue to seek advice on a topic in crystal optics; von Laue, who knew a lot about optics but little about crystals, elicited from Ewald the information that the distance apart of atoms or groups of atoms in crystals, though not known, must be of the order of a few Angstrom units. It was already accepted that *if* X-rays were waves they must have very short wave-lengths, and von Laue realised that they ought therefore to be diffracted by crystals. He got Friedrich and Knipping to carry out suitable experiments, which they did with great care, and so the first X-ray diffraction pictures were obtained. Von Laue's very complicated theoretical explanation may have delayed acceptance of the results, but their immense importance was soon realised generally.

WHB quickly appreciated the relevance of these experiments to what was then his main field of research. As soon as he saw von Laue's paper he tried to interpret it in terms of his own corpuscular theory of X-rays. He wrote to Schuster:

Rosehurst, Grosvenor Road, Leeds.
July 21

Dear Professor Schuster,

I enclose a drawing of the curious X-ray effect obtained by Dr Laue in Munich. It is claimed, I understand, that it is a diffraction effect due to the regular arrangement of the molecules in space. It is necessary that the crystal be cut pretty accurately to the axis: zinc blende and copper sulphate are the only two crystals mentioned in the letter I received.

I have got some zinc blende specimens and am going to

try the experiment: I have not yet found how exactly I am to cut what I have got, but I dare say I can manage that.

If you have any suggestions on the general question I should be very grateful. Of course it would be most interesting to try the penetrating powers of these side streams of rays. I am told that 'absorption by aluminium has shown that the rays producing the surrounding spots have a penetrating power very much greater than that of the main bulk of primary X-radiation and very much greater than that of the radiation characteristic of the anticathode.' I wonder whether the rays producing the side spots are really 'rays' proceeding in straight lines from some point in the crystal (say where the X-ray impinges or emerges), or are they sections of some loci by the photographic plate. It all seems most mysterious!

Sincerely yrs, W. H. Bragg

Later he wrote also to Rutherford:

I am awfully keen to know whether these reflected rays ionise, or at least produce cathode rays. Has Mosel[e]y found out? C. T. R. Wilson says, no; but his photographs are not numerous as yet. I am trying directly, & have got no sign of their ionising so far, as much as they ought to. I have just rigged the thing up better, for a more perfect try.

WHB's excitement was intense; the old Leeds troubles were dissolving. When WL came home for the summer vacation of 1912 WHB called his attention to the paper, suggesting that here was a line to be followed, a puzzle to be solved. They must have discussed it avidly: both men were thrilled by the prospect of a new series of experiments, and WL returned to Cambridge brooding on X-rays and crystals.

He wrote to his father:

. . .I want to have a good go at this crystal divining and I can now I have found the apparatus can be rigged up so simply. . . But don't let on for a bit till I can try the crystals Pope is getting for me. . .

Your very selfish son, WLB.

At Cambridge, WL had his 'brain wave' and solved the riddle. It was WL who saw that X-rays were to be explained at least partly in terms of waves and that Laue's photographic spots were wave-reflections from crystal planes; it was WL who adapted the optical diffraction equation and applied it to X-ray diffraction, where it became known as Bragg's law:

$$n\lambda = 2d \sin\theta$$

He had seen the way through the problem before his father and WHB was wholeheartedly delighted with his son's success. A postcard survives written by WL to his father late in 1912:

> I have just got a lovely series of reflections of the rays in mica plates with only a few minutes' exposure! Huge joy. I think the mirror will be a possibility. The spots vary but little in intensity over a large range of angles of incidence, of 4°–10°. WLB

And one can sense the pride with which WHB passed this on to Rutherford:

> Dec. 5 [1912]
>
> Rosehurst,
> Grosvenor Road,
> Leeds.
>
> My dear Rutherford,
>
> My boy has been getting beautiful X-ray reflections from mica sheets just as simple as the reflections of light in a mirror. They can be got in five minutes exposure, that is to say the principal reflected spot: the others don't come in so soon; but you get several pretty plain in 20!

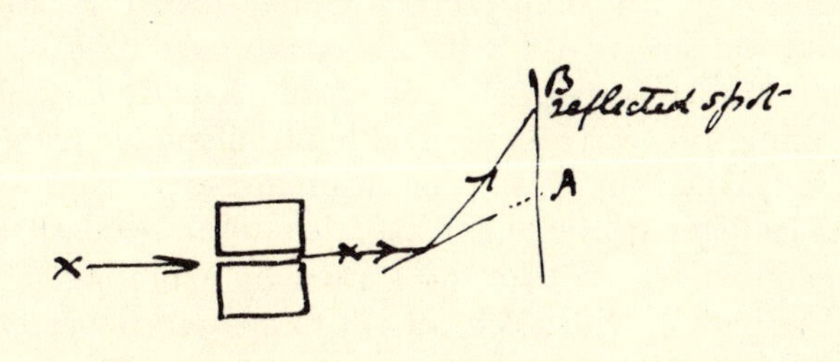

> Also you get a curious darkening close to A, lying between A and B; which he has not explained as yet. I have not heard from him for a few days but shall see him next week and his photos and will tell you how he is getting on. . .
>
> Good luck!
>
> Yrs W. H. Bragg

Fifty years later my brother wrote down his memory of this period in a letter to Professor Ewald in America who was preparing an account of WHB for his centenary year 1962. A shortened version of WL's letter is included in the Appendix, letter A.

The wave theory had won the day, in this instance; WHB acknowledged it at once; yet, as he wrote in *Nature* (28 November 1912), 'The properties of X-rays point clearly to a quasi-corpuscular theory, and certain properties of light can be similarly interpreted. The problem then becomes, it seems to me, not to decide between two theories of X-rays, but to find. . .one theory which possesses the capacity of both'. WHB was not going to discard the truth as he saw it, while fully acknowledging his son's results. He was content to wait in all humility, until, as it turned out, Einstein's 1905 work on light was resurrected and made clear (see pp. 60 and 127).

WHB wrote to Rutherford on 3 April 1913:

> I think I have got the whole thing more or less. It is against my corpuscular theory in part, but also against the pulse theory in part. However, that can be argued separately: I will just give you the facts. [He presented 'the facts', and closed with a postscript:]
> My boy and I have been working hard at the experiment lately.

Though WHB had been primarily interested in radiation, both father and son saw in WL's discovery the possibility of learning something fundamental not only about X-rays, but about all matter. While WL was fumbling with pitifully inadequate apparatus at the Cavendish, WHB – an inspired designer of apparatus, with his brilliant instrument maker Jenkinson – constructed his X-ray spectrometer. With it, he started by exploring further his study of radiation, while WL set out to analyse the arrangement of atoms within the crystal. But the two sides of the work soon merged into the new science of X-ray crystallography.

It was a marvellous marriage of two minds. WL provided the idea; WHB the means by which it could be tested and used. Had WHB not built his spectrometer, WL's idea – on which others were already working – might never have come to fruition as it did. And it was through WHB's expertise that the study of crystal structure leapt ahead in England, rather than in von Laue's Germany.

WL gave his father full credit for his skill. Speaking many years later at the Royal Institution, he said:

> My father was supreme at handling X-ray tubes and ionization chambers. You must find it hard to realise in these days what brutes X-ray tubes then were. One could not pass more than a milliampere through them for any length of time or the anti-cathodes got too hot. The discharge drove the gas into the walls; one then held a match under a little palladium tube which allowed some gas to diffuse through and so softened the tube. The measurement of ionization with a Wilson gold-leaf electroscope was quite an art too, and my father had thoroughly mastered all the techniques in his researches.[1]

Working together, 1912–13, results came tumbling out. WL described the time like looking for gold and finding nuggets lying around everywhere. Their most dramatic collaboration was in establishing the structure of the diamond.

They were so happy. For WL this was promise of a life of research; and for WHB the sun shone as it had not since Adelaide days. He was reinstated in his own eyes, with his position at Leeds justified after all: and he had the delight of collaborating with his own son, both at the height of their creative powers, the one having come to research so late, the other so early.

But then a tenuous cloud began to form; it was for WL first that the sun became obscured.

WL had published a guarded early speculation about the von Laue photographs in the *Proceedings of the Cambridge Philosophical Society* in November 1912. But this paper, called 'The Diffraction of Short Electromagnetic Waves by a Crystal', was not a strong statement, and WL was anxious not to refute altogether his father's corpuscular theory. The Journal was of rather limited circulation as well. He also wrote an article for *Science Progress*, entitled 'X-rays and Crystals', in which he flatly opposed his father's corpuscular theory. WL relied on these two articles to establish his priority in discovery; so they did, in print, but not necessarily in men's minds. WL had begged his father to keep quiet about their early projects, but WHB had of course discussed the results with his colleagues. The crystal work was directly relevant to his theory of X-rays; and he was open by nature. Perhaps, excited by the work, he hardly noticed at the beginning how praise was flowing in his direction. A letter from Lord Rayleigh illustrates the state of affairs:

31.10.12.

Dear Professor Bragg,

I am glad that you are giving attention to Laue & Co's spots and think you have an explanation wh. fits the facts. I did not know that you had a scientific son, and I congratulate you on it. I also am fortunate in that respect.

But WL longed for it to be recognised that it was in fact he who had had the first idea about those spots. Actually, WL's articles had not appeared when this letter was written; but when they did, they were neither strong enough nor important enough to counter the impression produced by the publication of WHB's two short notes on the subject in *Nature*. In both these notes, and in later articles as well, WHB made it quite clear that certain key ideas were his son's work, but such details are easily missed by readers interested only in the results. And what really made impact was the joint paper, ten pages long, written by WHB and WL and published in the *Proceedings of the Royal Society* early in 1913, setting out in as complete a form as possible the basic principles of the new subject. But it was joint; it was not WL's own.[2] A few months later, WL did write a paper of his own on the structure of sodium chloride, and this was followed by the joint paper on the diamond. But by this time the new line of research was being regarded as a mutual affair. It was not WL's especial province.

It was quite naturally the established professor who announced their results at the British Association Meeting and at the Solvay Conference on the Structure of Matter in 1913. The Solvay Conference Proceedings record WHB's full acknowledgement of WL as originator of the idea, and members of the conference sent a postcard to WL to congratulate him for 'advancing the course of natural science'. The card is signed by Sommerfeld, Curie, von Laue, Einstein, Lorentz, Rutherford and others. So *they* knew, but it is significant that in the summary of WHB's paper which appeared in the British Association Report, no mention is made of WL.

Honours, as allocated by the scientific world at large, were not easy – indeed very uneasy for WL. WL had written to WHB asking him to send those original papers of his 'to whoever would be interested'. He wrote another letter:

19th July, 1914

Dear Dad,

Who was it wanted copies of our papers, was it Lorentz? These ought to go, but I could not remember the name. You said you sent copies of all but my original paper. I should like to send him a copy of this for the sake of priority, for Ewald's paper on zinc blende in the Phys. Zeit. is all about that.

This shows WL struggling to establish his priority right up to the war. A young man has his name to make: it was natural.

Why had not WHB sent that original paper to 'whoever it was'? Had he just forgotten, or was it that, so excited about the work, he sent the later joint papers which explained it more fully than the first somewhat tentative one?

WL felt sore; hoping to enjoy a triumph, he felt it draining towards his father. He had been utterly at one with his father, sharing so much excitement. Now the world's approbation was damaging a relationship.

Could WHB have guarded against this situation developing? Perhaps, if he had been different – if he had been more worldly wise, less simply good. But his strangely solitary upbringing, his privateness, even his easy success in Adelaide, all prevented him, I believe, from knowing quite enough about how the world works. 'One mustn't be suspicious', was a favourite saying of WHB's – and though this trust was usually repaid and people loved him for it, on this occasion a little wisdom of the serpent would have served him better and might have saved a long heartache.

One could argue that WHB's imagination was somewhat slow to come into action: he always said his mind worked slowly. The 'unadventurousness' he blamed himself for at Cambridge, his late start on research – were they both examples of this? Perhaps another indication of this slowness was the extraordinary arrangement which WHB made for WL when he entered Adelaide University at the age of fifteen. WHB installed his son to work in his own room, and saw nothing odd in holding interviews and conducting college business in front of the boy. WL remembered the embarrassment and also how the plan cut him off from his fellow students (though of course they were not his contemporaries; he was much younger). The explanation may be as simple as that WHB felt it was the work that mattered most, and that WL would get on better in his room. Having less than the usual need of social contact himself, he may not have realised WL's social problem. WL endured it, but it seems a strange thing for WHB to have done.

In the years following their joint discoveries, it was natural for the senior man to be asked to lecture. Speaking in England and in America, WHB would hand credit to 'my boy', beaming with pride. But the audience would only go home thinking '*What* a nice man Professor Bragg is.' In a letter (April 1975) Sir James Cook recalled how, in his lectures at University College, WHB always referred to the wave-reflection theory as 'what my boy did', and I, too, remember how my father would 'lean over backwards' to give WL his due. But direct credit was what WL wanted.

It was difficult for the young WL; father and son never managed to discuss the situation, WHB being very reserved and WL inclined to bottle up his feelings. And WL had strong feelings – his mother had given him the dramatic and artistic nature which concentrates on a point, putting the rest of life somewhat out of focus, and judgement a bit off balance for the time being. He had been brought up, too, in her tradition of unselfishness at all costs; he felt things strongly but he could not hurt his father by telling him what he felt. However hard WHB tried to redeem a situation that had grown up almost inadvertently, and in spite of a shared Nobel Prize in 1915, the trouble lingered down the years. How they sorted out their lines of work after the 1914–18 war is described in Chapter 9.

6

WAR WORK AND LONDON UNIVERSITY: 1914–1923

August 1914. WHB was up in Cumberland the day war was declared, and used to talk of the strangeness of that day. He had gone there for the golden wedding of an uncle and aunt in a country vicarage; the old people were so taken up with their celebration that they took no notice of the war and he could not get a newspaper.

WL and RC were in Territorial Camp, WL being now a Fellow of Trinity College, Cambridge and RC an undergraduate, also at Trinity, with an Engineering Exhibition. They had joined the 'King Edward's Horse', originally called the 'King's Colonials', a troop started by colonials at the University. No doubt WL and RC were still feeling very Australian, so recently though permanently transplanted. RC was never to come out of uniform again. WL returned to Cambridge, applied for a commission, and was posted to the Leicestershire Royal Horse Artillery where he was a fish out of water among a lot of hunting men. But after a year's curious experience of men and horses the War Office plucked him out and sent him to France to take over the French method of locating enemy guns by sound, and start sound-ranging for the British Forces.[1]

I was seven years old in 1914, just old enough to feel an atmosphere and keep a strong if confused memory of bands and marching soldiers, flags and 'King and Country', of grave faces and wild excitement, and knitting, and Belgian refugees who teased our dog, and of polishing my brothers' Sam Browne belts.

In November 1914 WHB was in USA and Canada, lecturing on his and WL's joint work, a tour no doubt arranged long before war broke out. He wrote back to GB from Boston:

> The war news is very heavy still; though we seem to be holding the Germans well. The Americans are keen partisans of our side. They shake me warmly by the hand because I am English and follow everything keenly. . .they want to know what they can do to help. They show me their

knitting – the women do – and are said to be considering it right to knit in church. One man, a lawyer in Providence, came to me this evening to know if I could lend him some literature about the war, as he is lecturing on the Allies' side and has to meet a German in debate on Wednesday.

WHB curtailed the tour though he felt it had been 'thoroughly successful from a scientific point of view' and on one occasion 'it was very jolly, they were all so keen to hear about the things'. He wanted to be back. 'I have not got a berth yet but I think they are not hard to get now. If I go on the Lusitania I shall get home very quickly if all goes well.'

During the winter of 1914–15 WHB was thinking over an invitation to go to University College London. After exploring the research facilities there, he accepted; his main reason was to be nearer the centre of things where he hoped he could be of use in getting science and scientists employed to help the war effort.

He explained his reasons carefully in a letter to Arthur Smithells. Though the Smithells lived within a quarter of a mile of us in Leeds it is characteristic of WHB that he should have written; he always preferred to order his thoughts in writing rather than speak them. The letter to Professor Smithells is quoted later (p. 138). It is long, bearing on industrial and national affairs and so belongs more fitly to Chapter 10 than in the middle of a family narrative.

In the summer of 1915 WHB and GB came up to London to look for a house. RC had gone to Gallipoli with the Expeditionary Force. WHB wrote to RC on 22 August:

My dear old Bob,

Still at the Hotel York. Mother will have told you that Bill [WL] got his orders for France and crocked up the same day... He is particularly cross as you may imagine. I don't suppose it matters *very* much because the French people were really not ready for him and he was being sent as somewhat the result of his own importunity... We were delighted to have the letter you wrote when just leaving Alexandria: a letter from you just now is a great event. We have also got the one from the advanced base... We shall be sending you another parcel tomorrow.

When WL had recovered and been sent off to France, WHB and GB went back to Yorkshire, Mother to join Nanny and myself at Deerstones, my father to Leeds. But one morning as I was standing by the shallow stone sink in the kitchen, looking out into the garden, my father unexpectedly passed the window, came in, said to me quickly in a low voice 'Bob's gone' and went upstairs to my mother. I heard her cry out. All the rest of that day she walked up and down the flowery meadow by the cottage, a dark veil hiding her bent head.

It had happened to them as to so many parents. WHB's letter to Bob was returned with 'deceased' written on the envelope.

A letter to Rogers, written 30 December 1915:

> . . .I am getting on with the research work again. I see a lot of Professor Rutherford. He came over yesterday to discuss some new theories of his. We agree closely about things, and are almost in opposition to the Cambridge people.

The research work of 1913–14 had brought the joint award to father and son of the Nobel Prize for Physics in 1915 'pour les recherches sur les structures des cristaux au moyen des rayons de Roentgen'. WL got the news in France. The old curé on whom he was billeted got up a bottle of wine from his cellar to celebrate with. The Prize, and the sharing of it, was infinitely gratifying and encouraging; but WHB had no more time for his own research work after the move to London and University College; war work was claiming him. In his last letter to RC WHB had written: 'I get a few interesting little jobs in connection with the Board of Inventions, and hope to get more.'

One of the jobs was testing optical glass for the Government. Optical glass, which had been imported from Germany, suddenly had to be made in England cheaply and in quantity. Soon there was more important work for him.

In 1915 the Board of Inventions and Research (BIR) had been formed under the chairmanship of Admiral of the Fleet Lord Fisher 'to initiate, investigate and advise generally upon proposals in respect to the application of science and engineering to naval warfare'. There was a Central Committee consisting of Fisher, Sir J. J. Thomson, Sir Charles Parsons and Dr G. T. Beilby, and a much larger consulting panel of scientists and engineers, including Rutherford and WHB.

This was a historic departure. Individual scientists had been officially consulted before; Davy and Faraday on mining problems; Faraday was scientific adviser to Trinity House, concerned with the lights in lighthouses (which he saw changed in his time from oil lanterns to his new electricity); he also was called in to help the glass industry. But never before had a large-scale organization been mounted to bring science and scientists to the help of the nation. In the years before 1914 the Committee of Imperial Defence had failed to make a study of the mobilisation of industry and scientific research for war; the scale and effort required in total war had entirely escaped its attention.[2] England was caught unprepared. Scientists were protesting that they were not being used. H. G. Wells argued that the war would be a struggle in invention, and Fisher pointed out bluntly (in his letter to Balfour accepting chairmanship of the BIR) that:

> the War is going to be won by inventions. Eleven months of war have shown us simply as servile copyists of the Germans. When they have brought explosive shells into damnable prominence, then so have we. When they produced grenades for trench work, then so have we. . . Noxious gases made us send Professors to study German asphyxiation! German mines and submarines have walked ahead of us by leaps and bounds, although many years ago we were in a position of apparently unassailable superiority.[3]

But now the scientists were to be given their chance; they would be listened to grudgingly for a long time, but it was the beginning of a new relationship and a new collaboration. It was probably natural that the Senior Service was the proudest, the least co-operative at the start; whereas the newest, the Flying Corps, dependent on aeronautical research, was the most willing to accept the civilian scientists and engineers.

The work of the Panel was divided into six sections; Section II, to which WHB was appointed, was to deal with 'Submarines, mines, searchlights, wireless telegraphy and general electrical, electromagnetic, optical and acoustical subjects'; but as the submarine menace was acute and the Navy having little success in countering it, the primary object of Section II was to deal with the 'detection location and destruction of enemy submarines' – before Britain was starved out.

The BIR had a Secretariat attached 'to deal with the preliminary sifting of proposals from inventors and others and such secretarial and recording work as may be found necessary'. The sifting of proposals was no mean task; ideas poured in.

A naval research station had already been set up in the Firth of Forth on a little rocky thrift-covered promontory called Hawkcraig Point: it was near Rosyth, the naval base. The station was run by a naval scientist Commander C. P. Ryan RN. The BIR sent a couple of civilian scientists to join Ryan, one of whom was Dr A. B. Wood from Rutherford's brilliant team at Manchester. Wood discovered on arrival that Ryan had never heard of Rutherford, which indicates the level of scientific knowledge in the Navy at the time. However, Ryan had made a start on the job, working at the location of ships by underwater sound.

To quote from an account of this period that Wood wrote in later years:[4]

> At the time of our arrival at Hawkcraig the state of our knowledge of underwater sound propagation in the sea was very primitive indeed. . .[based on experiments on the velocity of sound under water in Lake Geneva in 1827 and of explosion waves at sea in 1889]. Any information likely to be useful in detecting submarines seemed to be lacking. I was impressed by Ryan's achievements in designing and making successful. . .hydrophones. . .[of various kinds. With them] large ships of Beatty's Rosyth Battle Squadron could be heard, under favourable weather conditions, at distances up to 10 or 12 miles in the Firth of Forth when proceeding out to sea. . .[But] most of [Ryan's] work was empirical. . . He knew little or nothing about the theory of sound or the possibilities of designing equipment which would indicate direction of sound. He had, however, gone so far as to fit submarines with pairs of hydrophones – one 'port' and the other 'starboard' – which would indicate to a reasonably intelligent operator the approximate bearing of another ship.

Such was the state of affairs at Hawkcraig when the subcommittee for Section II began to visit. Rutherford often came up with WHB and the Secretary, Sir Richard Paget, and in the spring of 1916 WHB, after only a few months at University College, was seconded to be Resident Director of Research at Hawkcraig. He arrived in May, bringing extra staff.

A. B. Wood became WHB's right-hand man; they had the same modesty and integrity, and worked comfortably together under the difficult conditions of Hawkcraig in 1916–17. This was the beginning of Wood's distinguished career as naval scientist.

A letter from WHB in London to GB at Eastbourne (where she had taken me to recover from bronchitis) describes WHB's plans for work in the north:

> 32 Ladbroke Sq. Wednesday
>
> My dearest Gwenny,
>
> I am here still as you see. There is a lot to do down here and not much in Scotland: so it seemed best to stay. . .I have got all my people here working, the only difficulty being the lack of instrument makers. . .I am ordering tools to go north and there are many other things so that I find plenty to do to keep all on the go. I went and saw Merz today [Charles Merz an electrical engineer and co-member of Section II] who was very cheerful and encouraging, exactly as I expected. He was refreshing because he thinks on a big scale. With my training I get into small panics when I take risks of spending money with no results. He is accustomed to spend thousands in that way: and as he says, the Admiralty is a bigger firm than any that ever he advises for, and can afford to spend far more. So I am going to send up some good tools there and lay out plans on big lines. Fisher has been asking about the photophone and whether I would go and see him. I sent word by James [Secretary, with Paget, to Section II] that all was going well.

WHB ended his letter with a nice little story:

> I had lunch at Shoolbred's today, when a lady accosted me as she was going out and when I was obviously floundering in my attempts to recognise her she said she was Miss Macpherson and asked me about everyone at home. I did not catch on even then, but I did my best and asked how they were at *her* home and where she was staying. We were very pleasant and cheery to each other, but then she said wasn't I Mr Smith of Wendover. So I said I was sorry, but I wasn't: and that rather checked the conversation: and she retired blushing to pay her bill. Wasn't that a bit of fun?

We let our new London house and moved north. Aberdour, the

little grey sea-side village near Hawkcraig, provided lodging for the scientists and Ryan's RNVR people.

The 'Wavy Navy' was a mixed collection of men with no proper naval training, often rather interesting and not very young. Ryan's lot included a London theatre-manager and Lieut Hamilton Harty, already a well-known musician (later conductor of the Hallé Orchestra); on occasion Ryan employed him to tap hydrophones with a little hammer, sorting them into pairs of low and high pitch. Tradition has it that once a piano was specially hired and Hamilton Harty detailed to manhandle it; but I think that story may be apocryphal.

There were jokes and frustrations; there was dread urgency behind the work. Beatty and Jellicoe visited Hawkcraig in May 1916; two weeks later Hawkcraig listened to the underwater sounds of Beatty's Battle Squadron going out; heard and saw the stragglers limping back after the battle of Jutland.

After WHB arrived, work concentrated on improving hydrophones: the original hydrophones could only tell that a ship was approaching, and they had to be refined to give knowledge of the direction of approach. 'At Hawkcraig [Wood wrote] we had laid down the basic principles in the design of a portable directional hydrophone (P.D.H.) and had made the first successful model. The work of making a reliable 'service type' P.D.H., either bi- or uni-directional, and to get these hydrophones into service as quickly as possible was therefore a high priority item', though not achieved until after they had left Hawkcraig.

There were delays and difficulties in getting the hydrophones fitted, but they were used in considerable numbers on drifters patrolling the coasts and on other small craft, and a few on submarines all through the war; they could 'hear' and give a pretty accurate bearing on a ship's propeller at several miles.

Many visitors came to Hawkcraig; members of the BIR Panel, of course, and Section II committee; Rutherford and Paget I remember at our house. Sir Richard Paget always enlivened things with his musical wit and ingenuity and unusual theories; he could sing a wordless two-part song on his own, accompanying himself on the piano, improvise an oratorio from an advertisement and would expound one of his theories such as that children should not be taught to read, only encouraged to make – I listened spellbound. But Allied men of science and importance visited

Hawkcraig too; there was exciting exchange; and some awful moments. Once a party of experts, French I think, were to come and inspect the work. Early on the morning they were to arrive WHB received a telegram from the Admiralty 'On no account show them anything' – and they were already on the way. WHB got in touch with Beatty's Flag Lieutenant who organised such a lunch on board the flagship that the party were in no condition to notice what they were *not* shown after lunch when they went round Hawkcraig Point.

Rumour was wild in those early years of the war, and ideas for winning the war were often wild.

'If the Board [BIR] had done nothing else', WHB said to a reporter 'it would have been worth while as serving as an umbrella to protect the naval officers from the deluge of letters... from the most hopeless cranks – from people who would try to catch a Zeppelin with a piece of limed string.'[5]

But sometimes the cranks did get their voices heard. There were holes in the Secretariat's net: perhaps the ingenious Paget even thought it would be fun to try the following idea (but this is entirely guessing). It was suggested to the BIR that sea-lions could be trained to hunt submarines instead of fish and so lead our motor-launches and destroyers towards the enemy. A.B. Wood got orders (not WHB's) to investigate this. With a sea-lion circus trainer trials were carried out in a swimming bath, in a loch, and finally, watched by an Admiral, in the sea off the Welsh coast: but the sea-lions did not come up to scratch and the full naval report concluded 'It is recommended that these animals should now be allowed to return to their legitimate business.'

These are a few of the stories from a time of great worry and tension for WHB. Personally, the scientific staff were on cordial relations with Commander Ryan and his dog (who persistently stole dinner off the table until they attached a high-tension battery to a steak one day); but over the work, relations were uneasy from the start, and became increasingly difficult. The Navy was not accustomed to co-operate with Science; and Science had to learn that the Services require a result that works every time and not *nearly* every time. It was a slow fumbling at the start. When scientific methods began to win over the trial and error methods of Ryan, it only made for more difficulty at Hawkcraig and with the Admiralty. There were jealousies and mistrust. A. B. Wood relates how one day 'when Prof Bragg was out in HMD *Heidra*, he asked the skipper of the ship to perform a certain manoeuvre. On the ship returning to Hawkcraig, Cdr Ryan sent for the skipper and gave him fourteen days CB (in the ship) for disobeying his orders. Prof Bragg went to see Ryan and apologised for having asked the man to do what Ryan had previously told him not to do and accepted all the blame, but the CB sentence had to stand'.

In the estuary there was a small island called Inchcolm of which the scientists made use in their experimenting. WHB persuaded Ryan to lay a spare length of hydrophone cable to be used as a telephone line between the island and the Point; but shortly after, in WHB's absence, the line was taken up again by one of Ryan's people without explanation, just 'by order'.

A state of strain had developed such that the scientists felt they could get no further with work at Hawkcraig. WHB explained the situation to Balfour who suggested that Ryan should be moved; but WHB said that as Ryan had been there first it would be better if the civilian scientists moved. It was decided to make a new research station at Parkeston Quay, Harwich.

WHB wrote a review of the state of affairs in a letter to J. J. Thomson (one of the Central Committee).

Dec 10 1916

> ...You expressed your regret, at the Panel meeting, at the delay due to the unfortunate situation at Hawkcraig. It is not perhaps so bad as you think. The work of the BIR has been the cause of useful advance in two directions. In the first place the BIR has pressed forward the plan of fitting listening apparatus to submarines which is now being adopted and used successfully. It is true that there has been much delay in fitting all the boats but the delay is mainly due to the reluctance of higher placed naval officers than those at Hawkcraig.
>
> The second advance has been the evolution of a practical directional finder which may be of real use. This was finished – in the main – last July: and again the delay in introducing it is due to officers at the Admiralty...[here WHB enlarged on the frustrations and difficulties of getting the instrument made]. Perhaps even if Ryan had been in full sympathy with BIR, we should not have succeeded

any better in bringing about the thorough trials of the instrument which ought to be made...

It is not so much to these things that the need for our transference to Harwich has arisen. It is more because we are not likely to get any further in a place where we are practically cut off from all contact with the navy, except such part of it as is hostile to BIR. I do not know that we should have got much further with the acoustic problem we were sent to Hawkcraig to investigate, if we had been entirely on our own: after all we have learnt a lot from Ryan. But there must be many practical developments of the problem, and *many other problems*, which we shall have a real chance of getting in touch with, in our new circumstances. I believe the direct contact between the men using the submarines and the physicists with their workmen will be really inspiring and fruitful. At Hawkcraig there were no submarine men, except the one officer on 13–3, who was of course not on really active service...

You may perhaps think that the break could have been made earlier than last month; but in the first place it would have been so immensely better if we could have brought co-operation about by persuasion than by making trouble at such a time, and in the second, when we had reluctantly come to the conclusion that we were not wanted we could only have brought vague charges, and if we had failed to make a case the situation would have been worse than before. We now have definite things to say, if required: I hope they won't be, nevertheless.

The move to Parkeston Quay took place early in 1917. It was a good site: the Quay where the trains used to arrive for the Hook of Holland boats had lain deserted since 1914; the Stour estuary by Harwich was the base for a destroyer and submarine flotilla.

With the rest of the staff's families, we settled ourselves in the small sea-side resort of Dovercourt Bay, near Harwich. Many of the little villas were occupied by naval families, jolly when the ships were in, but I remember the anxious restless feeling when a squadron was overdue.

The new Admiralty Research Station straddled the old railway lines along the Quay. There was space and greatly improved facilities for work; more and better ships: and WHB laid down that 'It is the duty of the senior officer or his deputy to carry out such requests [from the scientist] so long as he is satisfied that he can do so.' The old troubles they had had with Ryan were to be avoided.

With new opportunity ideas sprang to life. On 28 October 1917 they were able to borrow a 'big submarine'. WHB sent a scribbled note to Wood (who kept it) to say they would have the submarine 'all to ourselves to-morrow, for the day only, it goes away afterwards. So we must be quite ready: have all the apparatus seen to, and spares in case they are wanted! We have enough films? Get a couple more rolls if you can. You have everything else? We must see to stop-watches. I have seen Budgen. I shall be early... will you come early too?' WHB's excitement was infectious. Experiments bore results; solid progress was made. To quote A. B. Wood once again:

> It would certainly appear that at Parkeston Quay we 'skimmed the cream' off many of the submarine detection, location and destruction methods which have been used in two major wars.[6]

It was a terribly cold winter, 1916–17; a bitter east wind blew stinging clouds of dust and pellets of snow. Food was hard to get, with voluntary rationing; one potato a day for the household which was given to me as a skinny child; I remember spending a morning queuing for half a pound of margarine. Then there were air-raids. Early one sunny morning German two-seater planes swooped low over Dovercourt, low enough for me to see the face of an observer, bending down to look out. Nights were lively with barrage fire and searchlights. 'Will they come to-night?' I would ask my mother as I went to bed, and she would glance at the sky; raiders preferred fine nights and a moon. We had a dug-out in the back garden but it was too damp and cold, and the roof fell in before long. One night we watched a zeppelin caught in the searchlights burst into flames and crumple; it was the 'zep' that was brought down near Saxmundham, set on fire by the phosphorus bullets invented by Professor Threlfall; they pierced the skin and set fire to the hydrogen. This was the Professor Threlfall to whom WHB had written about physics in early Adelaide days, and always a friend. He also was on the BIR Panel. Threlfall was knighted for his invention, and GB with tact and difficulty prevented him from going to Buckingham Palace in tweed jacket and brown boots.

At first the work at Parkeston followed lines started at Hawkcraig, but soon it was branching out. Two gate-ships guarded the entrance to the estuary. It was found that hydrophones laid on the sea-bed between these two ships could give warning of approaching vessels; out of this grew experiments on an acoustic non-contact mine; WHB spent months at Mining School Portsmouth where trials of the new mine were being made during the winter of 1917–18. This mine, further developed by Wood, was widely and successfully used in the Second World War. Other useful developments were made at Parkeston, such as leader cables laid along the sea-bed to guide ships into harbour.

They were great occasions when WL arrived home on leave (always with a present from France for his young sister). Father and son would discuss their work. As a result, WHB suggested trying a sound-ranging system for the Navy comparable to the Army system. On land, the position of a gun was calculated by its pressure wave of firing being picked up by hot wire microphones spaced along a surveyed base-line; WHB's idea was that it should be possible to locate underwater explosions many miles away at sea by hydrophones arranged some miles off shore and connected by cable to a land base. This system was developed successfully, and has also been used with success in peacetime for charting the seas.

But the great achievement of the Parkeston Admiralty Research Station was ASDICS (Sonar), meaning 'Anti-Submarine Division – ics', a name devised to conceal what it was all about, which was the echo system of submarine detection. The principle of the system is that the presence of submerged objects can be detected by sonic rays directed towards and reflected from them.[4]

ASDICS could have warned the *Titanic* of the iceberg threatening it; indeed the basic idea was mooted at the time (1912) but not revived until the French scientist Langevin began experimenting in Paris in 1915. In 1916 the BIR took serious interest. Highly secret experiments became top priority at Parkeston. Trials had been carried out and ASDICS had reached the ship-fitting stage (the equipment was mounted in a stream-lined case let down through the bottom of the ship) when the war came to an end. And in the Second World War ASDICS came into its own, when WL was Scientific Adviser to the Admiralty.

We came back to London again when in 1918 WHB was transferred to the Admiralty as Scientific Adviser to the Director of the Anti-Submarine Division. By this time there was some grudging appreciation of the BIR's work. On 10 December 1917 WHB was able to write to A. B. Wood: 'I would have come down [to Portsmouth] before this but I have been so very busy with the [indicator] loop. It has gone surprisingly well, and will apparently work in very deep water, 50 fathoms at least... The authorities are getting a little excited about it.'

After the thrill of Armistice Day, WHB went back to his job at University College, London. Next year, 1919, came an invitation to move again. Sir Richard Threlfall wrote to ask WHB if he would like to go to Birmingham to fill the place of Sir Oliver Lodge who was about to retire, saying that 'the title of the office is being changed to Vice Chancellor'. WHB replied:

> ...I have been thinking hard of the proposal you made... But there is an insuperable objection... It would mean the end of my research work in pure science. I think there is still work in that field which I could do, though of course no one can be sure. Whether this is so or not, I ought to try: because for various reasons, of which the Nobel Prize is one, I am expected to go on with the research and should let down an ideal if I gave it up. [The Nobel Prize gave him confidence in this decision which he had lacked when he refused the suggestion to go to British Columbia seven years before.] It is understood that a University professor may shut himself up in his laboratory to some extent... But a Vice Chancellor must give his whole time to the work of his office... He could not have his mind full of research problems at the same time.

Merely as a Professor WHB was suffering under too much administrative work. He wrote to Sir Robert Hadfield on 20 November 1919 'My son and I have been comparing notes, [WL was by then at Manchester in the Chair vacated by Rutherford who had gone to the Cavendish] and we find we can only get a few hours each week for research.'

But although WHB was finding less time to experiment with his own hands he was gradually assembling a splendid team of research workers, supported by grants from the DSIR. Astbury and Shearer came first, then Müller and Miss Yardley (later Dame Kathleen Lonsdale) and others.

WHB was happy with his team as the X-ray and crystal work got under way, but it was a fight to get materials for their experiments, the hindrances 'serious enough to make good work impossible' as he wrote in a letter to Professor Andrade. Materials apparently had to be requisitioned far in advance of their need and it was quite impossible to get anything at short notice. WHB, used by this time to having the vast resources of the Admiralty and the BIR behind him, saw the obstructions of University College finances as petty and ridiculous. He campaigned against the restrictions, and this caused jealousy from those who were competing for funds. College politics depressed him.

WHB did not enjoy his time at UCL. However he was made KBE in 1920 and had the pleasure of making GB 'Lady'; and we all rejoiced over WL's marriage to Alice Hopkinson in 1921. In 1923 the time at UCL came to an end. Early that year the death occurred of Sir James Dewar, Director of the Royal Institution.

7

LIFE AT THE ROYAL INSTITUTION: 1923–1935

Sir Arthur Keith, a great anatomist, who with his wife became family friends of the Braggs, was Secretary of the Royal Institution. In his old age he wrote his autobiography: in it he describes the choosing of a successor to Sir James Dewar at the Institution. I quote his words:

> Dewar had hoped that Rutherford. . .would take his place. So one day when Rutherford had come from Cambridge to give a lecture at the Royal Institution we interviewed him. He explained that he was too deeply committed at Cambridge to think of changing. 'But', he said, 'I know of a man who is as well fitted as I am, or even better, to fill the billet.' We eagerly asked for the name of this man. 'William Bragg' he replied. I supposed he meant young 'William Bragg' (now Sir Lawrence Bragg). 'No,' he answered, 'I mean Sir William Bragg, Professor of Physics at University College, London: he is a great man of science and he is also a very great man.'[1]

It followed that on 7 May 1923 Sir James Crichton Browne, Treasurer of the Royal Institution, wrote to WHB to say the Managers had that afternoon decided to elect a successor to Sir James Dewar, and 'your name was mentioned in connexion with the vacancy and was received with *unanimous* approval'. Shortly afterwards Sir James, who wore Dundreary whiskers in Crimean fashion and had been a famous mental specialist in Queen Victoria's day, arrived at University College in frock coat and top hat 'to wait on' WHB. He waited until WHB returned from giving a lecture. Sir James then presented the Managers' formal invitation to become the new Resident Professor of the Royal Institution and Director of the Davy Faraday Laboratory.

WHB considered.

Sir James Dewar had died a very old man: vitality had ebbed from the Institution. Those who know and love the old place must forgive some description. It was founded in 1799 by Count Rumford, the volcanic American who had got his title of Count of the Holy Roman Empire by freeing Munich from beggars, vagrant soldiers who were a plague in Europe after the wars. He gave them shelter and work (they dug the 'English Garden'), and fed them from his scientifically designed soup kitchens. Delighted especially with the last, he returned to England where there was rising interest in philanthropy and concern for the working man. Mechanics' Institutes were being formed. Rumford turned his scientific mind to the improvement of trades and crafts, and conceived and drew up a 'Proposal for forming by subscription, in the Metropolis of the British Empire, a Public Institution for diffusing the knowledge and facilitating the general introduction of useful mechanical inventions and improvements, and for teaching by courses of philosophical lectures and experiments the application of science to the common purposes of life.'

The first meeting of subscribers was held in the house of Sir Joseph Banks, the President of the Royal Society; the following year the King granted a Royal Charter to the new 'Royal Institution of Great Britain'.

Rumford's idea was that workmen should come to study the latest scientific developments in their trades and listen to lectures (from the lecture room gallery only). His new inventions to improve domestic economy were on show, such as his cooking pots and the 'Rumford grate' (Jane Austen describes General Tilney standing in front of one in the drawing-room at Northanger Abbey). But the artisans did not flock to the Institution; money began to run out. Rumford threw in his hand and departed in a huff. His scheme had failed. But patient Thomas Bernard, Treasurer of the Royal Institution, carried on; the lecturing side became more important; the artisans disappeared. Humphry Davy was appointed: he was a brilliant lecturer. Davy turned the Royal Institution (known to many of us as the RI) into a meeting place of science and fashion; carriages blocked Albemarle Street when he was lecturing. And he started research at the Royal Institution. Faraday succeeded Davy, and was followed by Tyndall; the Institution had taken good root. The Davy Faraday Research Laboratory, founded by Ludwig Mond, was established in the next-door house in 1896.

WHB saw ample space and opportunity for his own research

in the Davy Faraday Laboratory; the Institution had a noble tradition in lecturing. My father had always worked for understanding between science and the other disciplines and saw the RI's potentialities. After making sure that his research grant from the DSIR would be transferred from University College to the RI he accepted the position.

We were acquainted with the RI because at Christmas 1920 WHB had given the Children's Lectures there on 'The World of Sound'. He used to convey myself and two friends to school of a morning, and that autumn as we walked along the Bayswater Road he would tell us about the lectures, and on Saturdays often took us to see the experiments that were being prepared at the RI: we all played with them. Once we were invited up to the Resident Professor's flat. I remember its dim richness and the Dewars' treasures, among them the pair of bellows made by Benvenuto Cellini. Faraday had been given 'four rooms with candles'; by this time there were fifteen odd rooms which ran the length of the building above the libraries, and although the flat was free to the Director the salary was modest, £1200 a year. If ambitious old Uncle William had not left him money 'to found a family' WHB could never have gone to the RI, he always said.

WHB took stock. The Davy Faraday Laboratory was almost deserted; two old gentlemen came occasionally, one of them we believed to be researching on spiders' eyes. WHB going the rounds found a store of platinum scattered in a dusty drawer.

The move began. WHB's secretary (Winifred Deighton) relates how she came in a taxi from University College with Miss Yardley (later Dame Kathleen Lonsdale), Dr Shearer and Dr Müller, each clutching his own fragile X-ray tube, and she (WD) holding WHB's tube. The famous X-ray spectrometer was installed. And we, the family, settled into the flat.

It was important to find out how the RI functioned. It seemed that habit was the only thing that had kept the place going, and that change would depend on the personal effort of the Director in every department.[2] At first even troubles with the cleaners came to WHB for settlement. Mr Young the secretary would come out of his little office rubbing his hands nervously; he had been scared of Sir James (who was a martinet); Mr Young still looked ready to hide. A Sandemanian, he was the last link with

Faraday. He wrote all the letters in longhand, there was no typewriter in his office; nor would he use a telephone.

Sometimes of an evening, in those early days, my father would say to me 'Shall we explore?' We would go down into the basement, past Dewar's huge old engines for liquefying gases, we would open the door still marked 'Servants Hall' (the RI had been made out of eighteenth-century terrace houses, with a pillared façade put on by Vulliamy in 1828). This door led into Faraday's laboratory and beside it was the 'froggery' where frogs were kept for electrical experiments; now a beautifully arranged museum, then the room smelt of old chemicals and dirt. We peered into the coal-cellars running under the street pavement piled with you-could-not-see-what under the cobwebs. From somewhere we unearthed a number of gold-and-white tea cups, a large gold mirror, a washstand with 'M.F.' on a brass plate and Faraday's table and chair which my father always used in his (and Faraday's) study in the flat.

The centre of the social life of the Institution was the Library on Friday evenings, the evening of the 'Discourse'. All week the Library had been heavy with silence unbroken by one white-bearded member who sat there reading in the same chair, all day and every day, and seemed part of the furnishing: but on Friday nights the Library was filled with members chatting decorously and examining exhibits. The Managers in white ties and tails stood grouped round the blazing fire with their ladies in long white gloves. Deaf old Sir James Crichton Browne would step forward – 'Ah, my dear young lady' was his invariable greeting to me.

One of the important jobs was to get to know the members. They had grown old alongside the Dewars, and there was a high proportion of old ladies who came regularly, no matter what the lecture was about. Once, at the end of a Discourse, the lady sitting next to me put her hand on my arm, 'My dear,' she said, 'I don't understand a word about this science, but I do love the sound of it'.

WHB took great trouble over the new lecture lists; he arranged lectures on a wide variety of subjects, getting the best people to talk on the latest work: and the Braggs started a new tradition of entertaining in the Director's flat. The Faradays had lived quietly upstairs, refusing invitations from London society; and Tyndall had lived a busy bachelor's existence until late in life, only his close friends ever saw the flat. Dewar had sometimes taken a visitor up for a glass of his special whisky: GB meant to do quite differently. Dinner·parties were instituted on Friday evenings when the outside world was invited to meet the lecturer and be introduced to the RI; and after the lecture, members who had never seen the fine old rooms in the flat were invited up. The parties were a great success. GB was a good hostess, and loved making a party 'go'. Of course it was no kind of 'salon' that she made; she had the wit but lacked knowledge: however she could swiftly divine when she was getting on the wrong tack in conversation and trim her sails: and most important, she gave the same warm welcome and interest to both humble and great.

The RI began to awake. On 10 April 1924, a year after our arrival, Sir James Crichton Browne sent a congratulatory letter to WHB. The membership had increased; 'The Institution is flourishing in every department [he wrote] and its only danger is hasty innovation which might derogate from its great reputation'.

Friday evening lectures were going well; but the afternoon lectures, arranged for 3 o'clock when ladies and gentlemen were traditionally leisured, were so poorly attended that one afternoon the choir for a musical lecture was larger than the audience listening to it. WHB suffered, and changed the hour to 5.30: his lists in the 'twenties include sets of lectures by Sickert, de la Mare, by Gustav Holst: but still, back from school, I would often hurry down in a gym tunic to make one more in the pitifully small audience. WHB tried the expedient of putting on courses of advanced scientific lectures and advertising them at London University, but although this increased audiences for a time, before long there were so many such lectures within the University that students did not need to come to Albemarle Street. The afternoon lectures were dropped after the Second World War.

The famous courses of Children's Lectures after Christmas, started by Faraday in 1826–7, were another matter; they were always given to packed houses. There is an old picture of the Prince Consort attending a lecture with two of his sons. Six lectures are given in the fortnight round Christmas to a 'Juvenile Auditory' which yet includes grown-ups and scientists come to admire the skill of experimental presentation and the discipline of explaining complex phenomena, and often new discovery, to school-age children. New inspiration has often sprung from such necessity. WHB in his time gave four courses, 'The World of Sound' already mentioned, 'Concerning the Nature of Things', 'Old Trades and New Knowledge' and 'The Universe of Light'. He took infinite care in their preparation; he spoke with a charm that drew his hearers into his own enthusiasm, for the job was near his heart, the job of showing wonder to the young (of all ages). There is a letter from Professor Smithells thanking him for one of the books that were made from these courses. He wrote (New Year's Eve 1924), 'Never grudge the time you spend on this kind of thing; the book you have sent me will do more for science than your modesty will enable you to estimate. But *I* know, and I am certain that your book will send out a wave of real intellectual, educational, and I would even say moral light and will help as much as anything that has happened to make the world a little wiser about modern science.' Appreciation was reassuring: receiving this WHB probably grunted, 'Ah, dear man'.

We became connoisseurs of lecturing as we listened to the Discourses, Friday after Friday evening. William Temple spoke in perfect flow, yet his notes were only a dozen words on a postcard; but G. M. Trevelyan *read* about the siege of Gibraltar without looking up, C. T. R. Wilson showed his famous cloud chamber experiment in an hour of misery for all, and Marconi – well, lecturing was not his forte. We always looked forward to lectures by Sir Leonard Woolley: he was a superb lecturer, finishing a rounded sentence as the clock struck: it was announced in the lecture list for June 1929 that Leonard Woolley would talk about his recent excavations at Ur of the Chaldees if he was back in time; the RI prided itself on providing the latest news of a subject. But the favourite lecturer of all was Rutherford: he coughed and boomed and his words exploded with enthusiasm and his lock of hair flapped and flopped down his forehead.

Rutherford, as a visiting Professor at the RI, was often in the Davy Faraday Laboratory, known to us as the DF. I remember the roars of laughter coming from the room where he was having tea and telling stories to the research people. WHB was so gentle and quiet; Rutherford so richly boisterous. 'It's a great thing, Life,' he once exclaimed, 'I wouldn't have missed it for anything.' WHB sometimes chuckled over the motor trip in the Pyrenees which he had shared with the Rutherfords in the early spring of 1912, and told us how their ears got so chapped driving in the open car that they bought raw lanoline from the mountain shepherds; and how Ernest and Mary argued over the map. 'I can never forget it all,' WHB wrote to Rutherford after they were back, 'the fun we have had and the many varied experiences. I never expected to go on such a trip and can hardly believe I have actually done so.'

From my earliest memory Rutherford was continually turning up at our home with an enthusiastic 'D'you know, Bragg. . .'

The Davy Faraday Laboratory, as I have said, was housed next to the Institution in 20 Albemarle Street, a dignified eighteenth-century town house. The RI had pushed through and taken over the first and second floors. The researchers worked mainly in the top floor rooms, many of them old servants' bedrooms, scarcely changed. Some extensions had been made at the back and in Dewar's time a hydraulic lift was installed, which was fun to work but required understanding.

Forming the original core of permanent research workers in the DF during the 'twenties were Shearer and Müller, Astbury, Gibbs, Plummer, Bernal, Miss Knaggs and Mrs Lonsdale; the combined monies paid to them, mostly by the Department of Scientific and Industrial Research, was under £2500 a year. But a number of people, from all over the world, would come to join the Laboratory for six months or so to learn the techniques of X-ray analysis. In the RI archives are two thick files of letters of application, between 1923 and 1928, and WHB's replies to them. The correspondence goes back and forth for months with a Russian who wanted to come; he was got over in the end. After their time in the DF these people would go out almost as missionaries to set up new laboratories for X-ray work in other centres and other lands.

There was often temporary work going on off the main stream; and so many requests came in from industry, even from archaeologists, for the investigation of different substances that at last WHB said that if the National Physical Laboratory would set up a department for the application of X-ray and crystal methods to such problems he would provide the man to run it. They did, and the DF lost Dr Shearer to them in 1927.

Although WHB managed gradually to assemble considerable funds there was not much money for running the laboratory in those early days; but the 'permanent' researchers were content with small salaries, the 'temporary' workers came on their own grants. The big and important pieces of apparatus were made by Jenkinson in the RI workshop, but the workers were expected to make and mend the small things themselves, there was no staff to help them. WHB also was quite willing for them to take on small outside jobs to help their pockets.

Those who were receiving money from the RI used to help a bit when WHB was giving a lecture or wanted something done. Miss Knaggs, a DF worker of that time, was telling me lately how they knew well that WHB liked things done at once (next week would not do). They (and she was one of them) would stay till late in the evening getting something ready, and next morning WHB would be so surprised and delighted. 'Those were wonderful days' she said.

Although he did not use it much, WHB loved to slip off to his own research room in the Davy Faraday where the spectrometer still presided, rather dusty. He liked to be able to turn to research with a free mind; before he went he would come and find GB to see if all was well with her. He put his head round the drawing-room door one afternoon and I heard him ask 'Is everything all right?' 'Yes,' said GB, but a certain aunt, known to be a stormy petrel, was coming to tea. 'Oh: is everything else all right?' – and he slipped away.

GB was always giving tea parties, to RI ladies, old friends, and of course to visiting Australians. Einstein's discoveries were causing popular excitement in the 'twenties: her friends asked her about them and she loved explaining; as she said, 'They understand so much better when I explain than when Will does.'

Sometimes WL would come and stay for a night, to give a lecture or for a meeting; things were going well for him at Manchester after a difficult start, when students were unsettled after the war and he was an inexperienced lecturer following after the brilliant Rutherford. But by the mid-twenties his research school was flourishing. His visits to the RI were looked forward to eagerly: but sometimes there was trouble, and myself the cause. My brother felt that I should have a chance for the University, but my mother was of the opinion that a little art and some parties would be better: WL would attack WHB who would groan and say he thought I kept my mother happy. Later WL would harangue GB, a hopeless approach, as she would then automatically talk to win on the other side; and next morning WL would depart feeling he had muffed it again. I felt thankful when peace returned.

Living at the RI – living 'over the shop' in such a strange old place with a constitution that ran only on good will – would have worn WHB down if we had not been able to escape to a country cottage.

The Wharfedale cottage had been the refuge and delight of Leeds days, but it was too far from London. We went back there once in 1918, driving up in our 1913 Austin car that had been laid up through the war: the tyres had rotted and WHB changed them on the roadside: roads that had not been mended for four years were so pot-holed that one enquired at each town for the best route to the next. Two nights on the way, we arrived on the third in a fog, with GB walking ahead carrying a detached acetylene headlamp to show the verge. That was our last visit. It was agony to leave Deerstones for good. For a couple of years we house-hunted at week-ends, and then acquired Watlands, near Chiddingfold, Surrey. It was romantic and old; never so lovable as Deerstones, the stream at the bottom of the wood was muddy; but it was within easy reach of London.

At the corner of the lawn was a wooden hut. It held jam pots and everything you didn't know what to do with, and apples in winter: we called it the 'Apple House'. It had a chair and a deal table in it, and here WHB worked in summer holidays. He could hear the voices when neighbours came to call, but remain unseen. Not that he was unfriendly, he just wanted to get on with his work: he thoroughly enjoyed a chat with the gardener and played billiards at the stockbroker's house next door.

Once a year all the Davy Faraday workers were invited down for the day. WHB hired a motor coach and GB prepared an enormous amount of food. 'All nice people are greedy' she would say, piling the plates. Croquet was played on the front lawn, bowls at the back. Some went for walks in the woods. Such a clever lot, pleased with the simple entertainment of the day. The sun seemed always to shine. Miss Knaggs has told me again how much they enjoyed themselves.

The plight of a neighbour across the road led to one interesting piece of applied science. Poor Mr Halahan had progressive muscular atrophy. For two years he had been kept alive by artificial respiration, applied by members of his family in relays. WHB conceived the idea of bandaging a football bladder to his chest and inflating it by means of a foot pump; the attendant could at least sit down. I remember WHB assembling the parts in the back garden and trying out his idea; the parts had naturally been prepared in the RI workshop. At first it worked clumsily; it worked better when the pump was connected to the water main. WHB enlisted the help of R. W. Paul of the Cambridge Instrument Company. Mr Paul entered into the problem with infinite kindness and friendship to the Halahan family, and the finished machine, known as the Bragg–Paul Pulsator, came to be used in hospitals. There is a description of it in the *Lancet*.[3] Mr Halahan lived with its help for another four years, alert and interested to his end.

That was some solution to a sad problem and beneficial; but I remember WHB getting dragged into another problem which distressed him greatly. During the years after the war, Spiritualism flourished. Bereaved parents, credulous and half credulous, incredulous, all ached for contact with lost children. Sir Oliver Lodge had lost his son; he was sure that he had got into touch with him again through a medium. Sir William Crookes was a firm believer. The Psychical Research Society worked earnestly, collecting evidence, testing, trying to secure proof. There was hope and doubt, fear of fraud: scientists were called in to check the genuineness of séances. WHB was kind; he was roped in by a noble lady to attend séances at her house. He hated it.

One evening at a dinner WHB found himself sitting next to Mr Devant, 'magician', of the famous Maskelyne & Devant Theatre where one watched an afternoon's worth of marvellous and breath-taking conjuring tricks in the Christmas holidays (ladies were sawn in half in boxes); the theatre was by the old Queen's Hall, alas for the disappearance of both! WHB must have told Mr Devant of his problem. Mr Devant observed that a 'magician' should be got in to sift the proceedings for fraud before the scientist made his tests. 'You scientists are the last people likely to detect fraud,' he explained, 'you are accustomed to deal with Nature who never cheats.'

It is hard to chronicle these years; WHB was working for the interests of the RI, to develop the Davy Faraday Laboratory; working for the DSIR to promote industrial research, and working for education, to get the province of science understood. This is described in more detail in Chapter 10. He opened laboratories, gave away prizes, received honorary degrees, and was becoming known as a broadcaster.

WHB began broadcast talks on science for the BBC in March 1924, soon after the foundation of the company.

John Reith had noted the success of WHB's Children's Lectures at the RI 'On the Nature of things' at Christmas 1923. Reith invited WHB to give a series of broadcast talks under that title. The BBC staff, as well as the public, were impressed by the success of the talks and WHB was invited to do another series 'On Sound' in the autumn of 1924. In January 1925 *The Observer* wrote:

'Sir William Bragg, whose voice is familiar to many thousands through the agency of wireless waves, possesses in an uncommon degree the power of putting in simple and graphic words the essence of a scientific theory, and rendering the imaginative and constructive aspect of modern science vivid to the man who lives outside the Laboratory'.

The BBC were keen to use WHB as the populariser of 'everyday science', but by 1926 some members of the programme staff wanted this sort of thing taken out of the main programme and transferred to the 'Schools' broadcasts. Policy was veering more towards entertainment and away from Reith's original vision of 'uplift'. WHB did not agree with the new policy; he thought science should have higher priority, with his longing for everybody to understand a little of what science was trying to say. But he remained on friendly terms with the BBC. He gave the 'National Lecture' on Faraday in 1931, the Introduction to Julian Huxley's series on 'Scientific Research and Social Needs'; spoke Obituaries on Mme Curie and Lodge, gave another set of talks, 'What is Light?' (following his Children's Lectures on 'The Universe of Light'), also many other talks including the one that sparked off 'Moral Rearmament' (see p. 110). His last great broadcasting effort was the organisation of the series 'Science lifts the Veil' in 1942; he died before the end of it.

But to get back to 1927; an incident pin-points the date. From *The Times* 30 December 1927:

'Two gas explosions, followed by fire, occurred in Albemarle Street, Piccadilly, last night, and caused damage to the Royal Institution. Fortunately there was no loss of life... The first explosion took place soon after 7 o'clock. There was a loud report ...and passers-by were alarmed at seeing sheets of flame from 12–15 feet high coming from one of the inspection pits or manholes in the pavement.' And so was I alarmed to see another manhole cover blown up in front of a second-floor window where I was dressing for a party: it was like an old picture of Vesuvius erupting. We were all rounded up (with difficulty because an elderly uncle on a visit could not find his glasses), told the basement was blown in and on fire, and herded into the lecture theatre: it was impossible to leave the building because a manhole was blazing in front of the only exit. Only a few hours earlier the theatre had been filled with a Christmas lecture crowd of children. It took the London Fire Brigade 'under Captain Firebrace' an hour to deal with the fire (and great credit should be given to two members of the RI staff, Green and Mitcham, who braved their way through dense smoke to turn off the main electricity switches); but before the hour was up we had been led across the road to Brown's Hotel where we sat eating sandwiches, all of us, from WHB to the kitchenmaid – the uncle *with* his glasses, and myself in a petticoat (under a coat). Meanly I went to that party later.[4]

Next morning the Libraries were a sad sight, windows shattered and everything covered with a thick layer of tarry soot. It was realised at once that major rebuilding of the RI was necessary: not only was there but the one exit; the body of the Lecture Theatre was found to be separated from the gallery escape stairs only by a thin wooden partition. Plans for rebuilding were initiated at once; it was the start of a long sore job not completed till 1930.[5]

A large proportion of the necessary money was raised by selling the 'Dorchester Papers', the correspondence of the American headquarters staff in the War of Independence, left to the RI by an early benefactor in 1805. The Papers included a number of letters from Washington himself. WHB got them valued, took a deep breath and doubled the amount to an interested American dealer, and got it. The purchaser later re-sold them for treble the price.[6] Poor WHB, it was not his kind of job; a question was asked in the House about letting the papers go out of England. It was an unhappy time; but for the family, a much worse time was upon us.

My mother was not well in 1928; she fell seriously ill at Easter 1929. Soon after, we had to move out of the flat for the rebuilding. She died in a furnished house in Chelsea at the end of September that year.

Our world was changed. She was so warm, so vital; if sometimes she wore her family down in her efforts to look after countless friends and lame dogs, we loved her and honoured the intent. GB and WHB were complement and contrast to each other; he so kind but detached, loving gentle cheerfulness with peace to get on with his work; she emotional and dramatic, always involved: he so reasonable, she working by confident intuition. I only once heard their voices raised with each other; he was protesting; she exclaimed 'But Will dear, why do you argue with me when I know I am right?'

GB had made and kept the friends for the family. When she died it was a fire quenched. I was twenty-two, and shivering had to take on the woman's job at the RI; WHB quietly got back to work. We managed; we were companions together.

Especially did I enjoy the companionship of going on trips with my father. Science offers delightful opportunities in conferences and lecture tours, pleasurable anyway for the unresponsible accompanying female. In 1924 I had been with my parents to the British Association meeting in Toronto, and we had travelled in the BA special train across Canada with stops for lectures and visits and receptions, and many meals of steak or chicken and hard-hearted lettuces. In 1930 WHB and I set off for the US; he was to lecture and I was so proudly his lady. WHB loved a sea voyage; perhaps it was the sea-faring blood in his veins. While I got anxious, his face smoothed out as the deck began to heave gently under our feet. He enjoyed the deck games, the slight social contacts of shipboard life, and he smacked his lips over the cold table in the dining saloon. We enjoyed the quiet on the top deck when the other passengers had already gone down to change for dinner and we watched the sun sinking down to a steel-blue sea.

Our last journey abroad together was in 1932, when WHB was sent out under the Prince of Wales' new scheme of Cultural Relations with South America. Lectures were to be given in Buenos Aires and Rio de Janeiro. We stayed at the British

Embassy in Buenos Aires. The Ambassadress was not much looking forward to having an 'absent-minded professor' to stay, especially with a daughter; the situation worsened when she asked us on arrival if we played bridge and we said no. But I think we turned out to be not so bad as she had feared. The University had not much standing in Buenos Aires, we found; it was the medical profession who commanded the position. But WHB's lectures were a success there, and at Rio. His lectures nearly always were, although he worried about them.

I was married soon after we got back, to Alban Caroe. To allow the newly-married couple to have a home of their own for a while 'Aunt Lorna' came to the rescue from Adelaide. She had always come over on visits to England whenever she could scrape together the fare. She came when my mother died. Now she came to look after my father and the household in the Royal Institution. But our connection with the RI was not severed for long, for after eighteen months we came back to inhabit part of the Director's huge flat. There we brought up our first two children, keeping a pram in the historic apparatus room; it was the first pram the RI had ever housed.

I think WHB was content; he was so undemanding, and my husband and I happily lived our own lives helping as we could with the entertaining and the life of the Institution and also sharing Watlands with WHB. He was a splendid grandfather. During the war years when their father was away, WHB and I brought up the children together.

Though my father was liked, loved and respected by many, his old shyness still remained. Once when I was going away, I suggested that he should invite someone in for the evening sometimes. 'But what should I talk to them about?' he said. Another time (I was away, just married) two of my aunts were keeping house for him; they were firmer. He was quite happy to read or to play patience with them of an evening, thankful not to be at a dinner making a speech (far worse ordeal to him than giving a lecture) but the aunts felt this was not quite right, that he should sometimes have company from outside; so they booked to go to a concert themselves and one of the RI managers was carefully selected to come for a nice dinner with WHB. But when they returned from their evening out and tip-toed up the passage past the drawing-room door they were disturbed at the complete silence; they peeped in. 'He went home early,' said WHB apologetically, looking up from his book.

Still, in the 'thirties, he began to walk down to the Athenaeum Club for lunch once a week; he could often encounter there someone with whom it was useful to discuss the RI and other problems. He came back one day, saying 'Do you know, the *Archbishop* came and spoke to me?'

Growing more and more involved in national and scientific affairs, he was drawn closer into contact with people; and with GB's protection withdrawn from him, he began to make relationships on his own. He could lecture with lucid charm, he could speak out when he felt strongly and talk with humble or great when he had anything to say; but clever talk alarmed him, and smutty wit worried him. William Nicholson painted a portrait of him which hangs at the RI. To keep his sitter amused Nicholson told him a stream of stories of the kind which WHB did not care for: the family maintain you can see this in the face portrayed. Trinity High Table conversation with its brilliance and College politics silenced him; however he enjoyed the quiet conversation of 'The Club'. He was elected in 1929 to the dining club founded by Joshua Reynolds and Dr Johnson, the very letter announcing election drafted by Gibbon. The members are drawn from different spheres; in WHB's time they included Stanley Baldwin, John Buchan, G. M. Trevelyan, Balfour, Hugh Cecil, both Chamberlains, Lord Crawford, Lord Davidson of Lambeth, Lord Sumner, Galsworthy and Kenyon (who had proposed WHB): they met at the Café Royal. Taking turns in the chair, they dined and conversed.

In 1930 the Royal Society bestowed on WHB the Copley Medal, its senior award, and in 1931 he was granted the Order of Merit. These awards came to him as votes of confidence in his work. The success of his bold planning for the Faraday Centenary in the latter year encouraged him further.

The Centenary celebrations were a tremendous task to organise: Thomas Martin, General Secretary to the RI, carried out plans with imaginative competence, and also edited Faraday's diary for publication in facsimile – the day-to-day record of his experimental research over 40 years. The Albert Hall was taken for an exhibition lasting a week. It was humbling to see the white statue of Faraday looking so small in the centre of the Albert Hall, with his own pieces of apparatus round his feet; the rest of the vast space filled with machines and technical devices of every description that stemmed from his work: all displayed anonymously, no names, no advertisements, with volunteers from Universities and Technical Institutions to explain them: all this in homage to the man at the centre.

Delegates from all over the world were invited and hospitality arranged; my own memories of functions are somewhat obscured by the difficulties of getting our Spanish guests there in time. There was a great Commemorative Meeting in the Queen's Hall. Sir Henry Wood conducted the BBC Symphony Orchestra playing Purcell's Trumpet Voluntary and Bach, the Prime Minister spoke, followed by de Broglie, Marconi, Elihu Thomson, Zeeman and Debye, backed on the platform by a row of notabilities including Smuts and Rutherford and J. J. Thomson. Then WHB delivered the Commemorative Oration, which was broadcast throughout the world.

We were glad to go to the country for the weekend.

In 1935 WHB was elected President of the Royal Society. For years he had dreamed of the lead which the Royal might give to Science and the country (see his 1915 letter to Smithells on page 138); now he was to have the chance to become its mouthpiece. Yet when invited to accept a nomination for Presidency he had hesitated. He wrote to the retiring President, Sir Frederick Gowland Hopkins on 10 May 1935:

> My dear President,
>
> Before I replied to your recent letter to me I have felt that I ought to consult my colleagues in the Royal Institution. I found them most helpful in that they would quite fall in with the suggestion of mine that I should abbreviate the number of my lectures, and they warmly encourage me to accept the invitation which you communicated to me. I have also thought it wise to consult my doctor for reasons which I need not go into very fully. [WHB was hampered by hardening of the arteries, he could only walk short distances]. This interview also resulted favourably. [The doctor said 'Better wear out than rust out'.]
>
> You will observe, however, that I am nearly seventy-three and that if I should hope to occupy the Presidential Chair

> for five years, which, I believe, is the normal period, I should then be seventy-eight. It would surely therefore be wise that the position should be reviewed at the end of two or three years.

In the event he stayed the five years' course, but he could never understand how he had reached such a position. He got very tired, but carried on determinedly. He would walk to Burlington House[7] from Albemarle Street with his slow old countryman's walk, the slight lurch copied perhaps from the men on his father's farm long ago; he would make several pauses on the way when he seemed to be taking deep interest in some shop window. One evening, opening his study door, I noticed how wearily he looked up from his papers. 'Daddy, need you work so hard?' I cried, and he answered simply, 'I must dear; I'm always afraid they'll find out how little I know.'

8

THE SHADOW OF WAR: 1932–1942

WHB was one of the most unpolitical of persons; he declined the invitation to stand for Parliament for the English Combined Universities in February 1937: but he felt strongly his responsibility to the Body Politic. To 'help your neighbour' comes again and again in his writings and sayings. In 1932 he was one of many signatories to a document called 'the National Memorial on the Disarmament Situation' presented to the Prime Minister;[1] in 1933 he was active on the Academic Assistance Council, for Hitler was then busy expelling Jewish academics from Germany. We had one refugee to stay for a while at the RI, a Jewish professor who had walked from Germany to the Black Sea. He was given new boots – they squeaked – and 2/6d a week pocket money while something was found for him to do. Imagine my feelings when he presented me with a bunch of tulips one day.

In *The Times* of 21 April 1934, there appeared a letter appealing for support for the Jews in Germany to be forthcoming from Christians as well as from British Jewry, signed Cosmo Cantuar, Cecil and WHB.

The horizon was growing darker through the 'thirties. In August 1938 came the Munich crisis. Early in September several Members of Parliament wrote to *The Times* saying security could only be achieved through moral regeneration. Then on 10 September *The Times* printed a letter over the names of Baldwin, Salisbury, Sankey, more lords and leaders and chiefs of the Armed Forces, and others including WHB. The letter asked 'To what is the world heading? What is the future of civilization? The world cannot for ever continue plunging from crisis to crisis. We must act, before crisis ends in catastrophe. The real need of the day is therefore moral and spiritual rearmament.'

WHB broadcast on this theme. 'I have been asked, as one of the signatories, to say what the letter meant to me.' He did, and it is worth quoting here most of what he said, because it expresses the essence of his attitude to politics and his country, as well as his creed.

> We are passing through very anxious times. One moment of dreadful danger has just gone by. . . Yet we cannot rest because we fear that other such moments may be in store. We suspect that if we had been wiser in past years the difficulties which caused the crisis might not have arisen, nor would there now be anything to deplore. What, then, can we do to gain rest in the future from our fears, to bring permanent peace, and to set us free for the kindlier affairs of the world?
>
> *The Times* letter asserted that 'the strength of a nation consists in the vitality of her principles. Policy, foreign as well as domestic, is for every nation ultimately determined by the character of her people and the inspiration of her leaders.' If we are disappointed in ourselves, we must suppose either that our principles are incorrect or that we have failed to obey them and are behindhand in the development of our character. What should be the principles of a nation? They can be no other than those which have been proclaimed by every great teacher of whom we have heard. Christ stated them in the form of two Commandments: 'Thou shalt love the Lord thy God with all thy heart', and 'Thou shalt love thy neighbour as thyself.' Socrates put it that a man must search for the right life to live, and then for the right way to live it; and his teachings implied that a man's life must be lived in relation to his fellow-men. . .'I spend my whole life,' said Socrates, 'in persuading you all to give your first and chiefest cares to the perfection of your souls, and not till you have done that to think of your bodies, or your wealth.' Later came the words of Christ: 'What shall it profit a man if he gain the whole world and lose his soul?'. . .
>
> Let us take this for granted, that we would lift ourselves, that we would, as many would say, approach nearer to God. We remember then that we have been told how to set about it: indeed, we somehow know already, but fall short of what our knowledge implies. The great teachers have insisted, and especially the Founder of the Christian religion, that we shall realise our wish if we can follow the second great

Commandment, namely that we shall love our neighbours as ourselves... It is the complete and sufficient complement to the first great Commandment. We may rightly speak, I think, of *trying* this way: it is an experiment that we are making, the greatest of all experiments. If we follow it, and can verify what we have been told, and have, indeed already assumed, each one of us, more or less, we gain confidence as we go. But we shall never be sure until we try.

I am here using language which comes naturally to anyone who has been in touch with that growth of natural knowledge which we call science, and acquire by experiment. This science does not, so far as I can see, affect in the slightest degree the great principles of which I am speaking. But it does affect the attitude in which we set out to follow them, and it can be most useful, and indeed necessary, in the attempt. Let me try to explain. The knowledge of Nature has greatly increased, especially in the last few hundred years, and has altered the appearance of the world and the lives of men...[But] it is important to observe that the richer acquaintance with Nature shows only an extension of the world which we knew before: the field is wider, but the character is the same. We do not proceed, so far at least, from the material to the spiritual. Scientific instruments do not bring us nearer to God in new ways. On the other hand, science can give us great help when we work for our neighbours, which is God's work...

All our [scientific] 'laws' so called, are hypotheses: assumptions that we make for the moment, and then put to the test. We keep only that which is proved, and even then we know that our proving may be inadequate because we have so much more to learn. Not only do statements need amendment from time to time because new experiment shows the necessity, but also because words change their meanings as the years go by. Indeed, they may not mean the same thing at any one time to the different people that hear them... So we perforce acquire a very sound humility in making our assertions. All of them are provisional: the only virtue that any hypothesis possesses lies in its ability to suggest further experiment; it is no use unless it leads to action.

It is natural for the scientist to express himself in similar terms when he speaks of the spiritual life. And not only the scientist, but numbers of men who have perhaps become the more inclined to do so because the scientific manner of thought and speech has become more usual. What one man calls a creed, another calls a hypothesis and means the same thing. I am sure that very many have found here a stumbling block. They have thought that unless they could say they were convinced of the truth of certain statements no progress could be made in spiritual life: they were outside the pale. But if they understand that there is a noble experiment to be made which will tax every energy of body and soul, and that they can wait for the results of the experiment to test the hypothesis which suggested it, they will respond. . .

The experiment may be very difficult indeed. Sometimes it is easy, as when we give some willing help that is obviously required; sometimes all the wisdom of the ages is insufficient to tell us what to do, as may be the case when our neighbour is faced with a sad distress or a difficult choice; sometimes we know too little about natural laws, as when some disease of body or mind is beyond our present science. Indeed, every effort to which love can prompt us, and all that we know, whether handed down from the past or revealed to us by modern search, is wanted for our task. Lazy kindness may be merely a mischief: we want far more than that.

What must we do to make the experiment? To begin with, we have to make ourselves fit for the work, fit in body and mind and brain. . . Fitness for what? Fitness to help one's neighbour. So also the mind must be cared for: it is not to be fouled by taking pleasure in the contemplation of foul things. It is to be trained to love good news of men, and to look for that goodwill, possessed at least in some measure by everyone, which we rely upon as the binding force of the community. . . Lastly, we must hold learning in high respect, because without it we cannot give the help that we would. . .

Whether our lately-won relief is permanent as we hope, or only a respite as some believe, now is the time to pull ourselves together. We see the efficiency of the authoritarian state, and realise once again the power of discipline and a common purpose. Only, in our case, we wish that the submission to discipline, though it may be equally severe, shall be rendered with intelligence and goodwill. Our service is voluntary, not forced; but it must be as full and as ready as if it had been ordered by a dictator. This is no more than a whole-hearted attempt to fulfil the second great Commandment. In this country we have the opportunities to make our effort while we are protected by the defences which we have been and still are constructing. The nearer we come to internal reconcilement, the more effective we shall be in arranging a reconcilement which is external and international; and the nearer we shall be to the day when all the defences can come down.

We do not know how far away this day may be. We can only be sure that it will not come unless we work and fight for it with all the devotion that a war would have required. It is for this that moral rearmament is necessary.[2]

Lord Kennet and Sir Walter Moberly continued this series of broadcasts. These talks raised tremendous response in a world that had been terrified, and then released uneasily at Munich from immediate terror. Voices from the Queen of Holland to the Spiritualists exclaimed in support, and the morning after his broadcast, while we were still at breakfast, two ardent young men called on WHB to say that their boss, Dr Buchman, American leader of the evangelistic 'Oxford Group', earnestly desired a meeting with him; could we come to tea (I was included in the invitation) that very afternoon at Dr B's headquarters at Brown's Hotel? WHB agreed a little cautiously. The 'Oxford Group' had considerable influence and following at that time, the vigour of a fresh missionary drive; several of our respected and intelligent friends were completely caught up in the movement – and yet...

That afternoon we crossed the street to Brown's and were shown up to a stuffy private sitting room where Dr Buchman waited, unimpressive behind spectacles, surrounded by disciples: and I was asked to pour out the tea.

Dr B was out to absorb WHB as an ally in his Group; while we sipped, a disciple with the fire of an Old Testament prophet proclaimed how some factory in the East End had been 'changed'; others backed up the evidence. I do not remember Dr B saying much; his was the role of conductor. WHB came away feeling Dr B's was not his way; the Group had failed to capture him.

But Dr Buchman took possession of the phrase 'Moral Rearmament', to WHB's distress, and used it ever after as motto for his campaign.

By the end of 1938 the war seemed close: however WHB left no stone unturned that he could handle in the effort to keep contact with German science. In his Royal Society Presidential address in December 1938 he announced:

> As the result of a discussion between Dr Krüss, head of the Prussian State Library in Berlin, and myself, when we attended the International Documentation Conference at Oxford in September last – a discussion initiated by Dr Krüss – I have lately received a letter from Dr Bosch, President of the Kaiser Wilhelm Gesellschaft, inviting the co-operation of the Royal Society in some scientific enterprise which would advance science and, at the same time, promote understanding and good will. Dr Bosch asks that a few representatives from this side should be the guests of the Kaiser Wilhelm Gesellschaft for a week during this winter and hopes that we would play the part of hosts in return. . .I feel sure that this gesture of friendliness will receive a warm welcome from Fellows of the Society.

Next month, in January 1939, a message of appeal to goodwill was broadcast by the BBC to Germany, in German. WHB was one of the eighteen signatories. This broadcast of hope against hope is printed in the Appendix at C.

In March the Royal Society was compiling a National Register of Scientists in preparation for a war effort.

In April WHB went to Washington to deliver the Pilgrim Lecture at the National Academy of Sciences, 'History in the Archives of the Royal Society'. Close links must be kept with the US. The President of the Academy thanked him, writing on 30 June 1939, '. . .In its own way, in the field of science, your visit was quite as much a success as the following visit of the King and Queen of your Empire'. An encouraging letter to receive.

In March–April two English scientists, Professors Donnan and A. J. Clark with Professor Dover Wilson, the Shakespearean scholar went to Germany on the exchange invitation that WHB had announced to the Royal Society. Anderson shelters were being made in May.

In June four Germans returned the English visit and came as guests of the RS, lecturing at the Royal Institution.

In July the shops were filling with black-out material and evacuation plans were perfected. During that hot August of 1939 women were all making black-out curtains; my sewing machine seemed to cover miles as I made curtains for the RI windows, wondering if the war or my third baby would arrive first. Then, one evening, WL telephoned to say we had better leave for the country at once, cars might be requisitioned next day: so, packing the back of the car with blankets for evacuees and the children's winter clothes, we left that night. For the next five years Watlands was the family home.

War was declared the next week, when my baby was four days old.

A letter from WHB to GB's sister Lorna in Adelaide gives a description of those first weeks of the war.

Watlands, Sept. 24, 1939

My dear Lorna,

This week I had your short troubled letter written after war had begun. . .

London braced itself for a terrific set of air raids, which have not materialized: the result is as curious as the non-happening. The Government find it hard to persuade people that the precautions were necessary: and certainly they were severe. London at night is as black as the country: people crawl about, and are advised to carry a newspaper or something white. Anyone who let a shaft of light out into the darkness was dropped on. The interference with business is terrific: I dare say a lot of the business was superfluous, people can live much simpler lives, but it is difficult for those who cater for the supply of luxuries, and have depended on the assumption that the luxuries were necessary. I expect you hear or read about these things...

I have a meeting of the RI Managers tomorrow. We have to settle our plans: the difficulty being again the uncertainty as to how far air raids will develop: & darkness must be maintained in the streets. The young people have largely gone to war service of one kind or another: the old people cannot get about. No Friday evenings at present, it is clear: nor even will the afternoon lectures have much chance of being useful. Some of the University Colleges have moved away from London: King's College has gone to Bristol, & University College to Birmingham, I think. I have been thinking that we may at the RI do something to help students who could not follow their Universities so far: in fact I have already made an informal offer...[3] Petrol rationing begins today. I think we get 7 gallons a month which will allow about four journeys to the station per week. I made a little store of petrol before the war began... I might apply for extra but I doubt whether I should get it. If I were given some special war duty I could manage to get something more: and I have some hope that I may get a mild job of some sort.

The evacuation has been a great movement, carried out most skilfully, but of course it opens up a great variety of new problems. The children without parents are in most cases doing very well, and I think they will profit by the move. The infants & parents are the difficulty. The mothers cannot stand the quiet, the absence of shops especially for fish & chips: and a lot have gone back, to the considerable concern of the authorities, who almost wish for an air raid to persuade the people that the precautions were not unnecessary...

I have bought a cinema projector: the idea being to hire films from the Kodak library & have a weekly show for the people & children round here who cannot now get to Godalming [our nearest town]. We had a family show last night & it worked well...

WHB spent the week at the RI, in company with his secretary Winifred Deighton, Ethel our cook, my husband and several other men whose wives and families had been evacuated. On Friday evenings WHB returned to Watlands with the week's film; and on Saturday evenings our small sitting room was packed with neighbours and their evacuees, hot and stuffy and delighted to watch old 'wild wests' and the adventures of Mickey Mouse on the small screen.

Fifteen months later WHB was writing to Lorna on 5 January 1941:

> There is such a lot I might talk about: but I doubt if it would pass the censor. So I am very careful. I am busy with various things. My five years' Presidency of the Royal Society is over: I can hardly believe it: but I am still Chairman of the Scientific Advisory Committee on Food, and on Hankey's Scientific Advisory Committee for the Cabinet. And there are other things. [Some account of WHB's work for National Affairs is given in Chapter 10; related here it would burst the bounds of this domestic narrative.] Only one weekly lecture at the RI now. I sleep there in a ground floor room half the week: the rest here [Watlands]. Only broken windows so far. [The Blitz was in full swing.] We have about 50 to 70 people sleeping regularly in the basement: most of them seem to like it. It is warm and dry, and there is company & music etc. Malcolm McGougan sleeps downstairs also & Miss Deighton & Ethel [our cook] who keeps house for us. . .Bosanquet has gone: his wife has come back from the States & they have a flat. He is being replaced by one of our other friends; a man called Cockcroft. You may have heard of him in connection with Rutherford's atom splitting. [They had beds ranged in the Museum corridor behind the lecture theatre, to be away from windows and flying glass. One light switch controlled all the lights, the man in the bed nearest the switch giving a last night call, 'Ethel, can I put out the lights?']. . . It is very quiet tonight: there have been several 'alerts' during the day, and a few bombs have dropped somewhere in London, I suppose. The guns were firing hard once or twice, but there has been no trouble near here. . . After lunch I sat with some other RS people on a panel which advises the Home Office on the desirability of various refugees to be released from internment. Later on I attended a meeting held in our theatre at which the fire-wardens of this district

> discussed their future arrangements. . .I was much interested in the views expressed by the different lots of people, business people, private people, landlords, tenants, people who kept in town during the week-end, others who did not. There is a wonderful lot of patriotic effort; and an irritating minority who won't do anything for themselves. On the whole it is good. The head of the wardens in this part – Mayfair – is an excellent person.[4] The shelters present a very difficult problem as you will have seen in the papers. Our own is much above the average I think.

The RI shelter had its regulars, with their own mattresses in their own spaces. Some inmates came across London to sleep there, and said it was the best. Among them were ladies who in those days waited for hire at the local street corners. At Hallowe'en the shelter inhabitants gave a party for all the RI staff with individual invitations; WHB's invitation survives, decorated with witches and spiders' webs by a Czech artist: there was food and plenty of drink and they all played Bob-apple. WHB goes on:

> I think you are right in supposing that the strangeness of our present lives is less felt by ourselves than by you. So much that is regular and ordinary goes on all the time that we are not thrown right off our ordinary feelings. No doubt people get hardened. At one time when the warning was given people trooped hastily into our shelter. . .today I noticed that though there was a 'red warning', i.e. a danger of greater imminence, most people took no notice. It was a very misty cold winter day, bitter cold: and I suppose that people realized that no airman could possibly know where he was and could only drop bombs at random, & the chance of being hit by a solitary bomb from a single raider was so very small that it could be risked.
>
> I am glad that the world applauds this country. It is terribly gratifying. Our men have certainly done great things, wonderful things: as good as anything in the past. I am glad the Australians have had their own special show. . .
>
> They have (at Watlands) a fancy dress party this week for the evacuee children! Aren't they energetic?

Life there in the country was busy, with billeting, knitting parties, coping with rations, dashing off on a bicycle at the rumour of *fish*; with small children and whooping cough and noisy nights. 'Haven't we had a lovely time' WHB's grandson said after two hours of reading the 'Water Babies' and eating ginger biscuits around midnight while guns banged and bombs were jettisoned in the woods.

Another letter from WHB to Lorna; at the head is scribbled

> Your parcel has just arrived intact! Hooray!
>
> June 23 1941
>
> . . . I spend long week-ends at Watlands now, generally catching an early afternoon train [on Friday] and going back early on Tuesday. The trains have been running regularly for some time. [In the worst of the Blitz it often took four to five hours to do the usual hour's run from Witley to Waterloo; the Portsmouth line was a favourite target, and parts of the journey often had to be done by bus.] The railways have done wonders in getting their trains under way again after damage: of which there has been a great deal. [But it was very exhausting for an old man with a bad heart, who could not walk much.]
>
> Yesterday we had news of the German invasion of Russia and we listened raptly to Churchill's broadcast last night: we thought it fine. . . Events are so big that one almost loses sense of their magnitude. I wonder how the Russians will get on. Their first communiqué came this morning: but it will take a few days to know how things are likely to develope. One of the wonderful things of this war is the presence of such good men in leading positions, Churchill, Menzies, Smuts, Fraser, Roosevelt, Cordell Hull, Winant: and the list could be longer. . .
>
> What a business this flying is! The progress made is staggering, and the proposals for enlargement of the air force take one's breath away. The preparations of the young men for it start early, according to published reports. I imagine there must be one or two hundred thousand youths enrolled in the new organization for boys 16–18. I rather think that lectures will be given for picked members at the RI: and it may well be that I shall have a shot at giving one or more of them myself.

WHB did give a lecture to the Air Training Corps at the RI,

three times repeated, which was published in December 1941 under the title 'The Story of Electromagnetism'.

The long weekends at Watlands were relief after the stress of London; later in the same letter he writes:

> I am sitting in the apple-house this morning: it is lovely & cool after the stuffy heat of the last few days. I am pottering away at some crystal calculations: I have sent two papers to the Royal Society lately, not claiming to do a great deal, but quite interesting I think.[5] Of course the younger generation are far ahead of people like myself, but I am not sure but that this time they have got so far ahead that they have missed something. They are sitting on me rather; but I am content to wait. The Americans have nothing to interrupt them and get on fast: this new business of crystal experiments & theory is a favourite of some of them. So I am quite pleased that my staff at the RI, women and old people and refugees, are doing something to map out new ground. I do a lot of calculating, rather like knitting for a woman; I have the time because although I preside over various committees I do not have much actual work to do. I have rather more than dear old Sir Dundas Grant who says he finds he is wonderfully useful at 'making' quorums! He is well past 80: I enter my 80th year myself on Wednesday week!

There is one more letter from WHB to Lorna, written on 25 January 1942, from Watlands. It starts:

> A very welcome present of marmalade and sugar was waiting here for me when I came down this week-end. It is a great pleasure to the household when something extra arrives, something outside the rations. Not that we are short of food. [By 1942 we were used to living on 1 egg a week, ¼lb of butter and 1/- worth of meat etc.] It is marvellous how the supply has kept up. [He goes on to describe several new devices for keeping up supply, such as de-hydration]. . . Carrots are in little shreds; we like to chew them like sweets!

Later after much family news he writes:

> I have had quite a lot to do with the broadcasting lately. I am Chairman of the committee of the British Council which

> deals with the propaganda of science abroad. BBC has arranged a series of twelve talks on science matters, the subject is 'Science lifts the veil' which is a nice comprehensive title. I gave the first myself and have to superintend the rest[6]. . .I am also giving two school broadcasts, and a Home [Service] talk on the Royal Institution. Also helping the British Council people to make a film of the RI. I am acting as President of the RS again while Dale is in America; and there are other committees, so my hands are quite full. Our lectures go on at the RI, we get fair audiences; people are glad to listen to something other than war news. Just lately we have only had fifty or so but the snow and frost have been severe, people could not get out. Today is fine and sunny and the snow has nearly gone.

But the slush froze again, the snow lay freezing and melting, growing dirtier, from Christmas to March that year. WHB had to go up on a Monday from Watlands; the broadcast series 'Science lifts the Veil' were on Monday nights. The theme was his, that of showing the place of mankind in the range of sizes in the Universe, for, as he said in his introductory talk, 'We do indeed live blindly, unaware of the greater world, perceiving only a small portion which we are apt to think of as all that matters. One of the first consequences of the discovery of the microscope was to wrench open men's minds, and to let in the light of day on their self-sufficiency; to lift a veil which hid the view of a greater world.'

Each Monday WHB introduced the speaker: he introduced ten, including WL.

The talk by J. D. Bernal on Monday 12 March was in the form of a dialogue with WHB on 'The Problem of the Origin of Life.' WHB was so very tired; his heart could struggle no longer. On the Thursday after that last talk he died.

His memorial service was in Westminster Abbey – the first to be held there since the bomb fell through the Lantern, so we were told.

9

W. H. BRAGG AND CRYSTAL ANALYSIS: 1919–1942

I have only written about this period of WHB's life, 1919–1942, from the family angle; his death brings a domestic narrative to a close. Some review must be made of WHB's specifically scientific work during this latter part of his life, and for this I must rely on quoting from scientists who knew him and knew the work as it evolved during the time; remember, I am no scientist myself. My sources are the finely written Obituary Notice that Professor E. N. da C. Andrade wrote for the Royal Society,[1] and a little account in *Nature*, written by Professor Astbury who knew WHB through day-to-day encounter when he worked as one of his research team.[2]

Professor Andrade wrote as follows, describing the period when WHB was getting back to academic work again after the First World War:

> In those days Bragg was himself actively experimenting with his own hands, as well as directing research: he was seated at his spectrometer whenever he got the chance. At the beginning of his work at University College he still used his first well-tried weapon, the ionization chamber, as detector, but gradually gave it up, for most purposes, in favour of the photographic plate. The equipment at University College was somewhat scanty at first and Bragg and his students, Müller and Shearer in particular, set to work to develop it. Continuously evacuated X-ray tubes, both hot wire and gas filled, were introduced and a self-rectifying gas tube was evolved, which gave useful service for many years. In those days vacuum pumps had not reached their present state of efficiency and hot wire tubes were rather troublesome. The work was supported by generous grants from the Department of Scientific and Industrial Research, which were more than justified by the results.
>
> The University College period was notable for the first attack on the structure of organic crystals. For much of

this work Bragg employed the powder method, whereas hitherto he had worked with single crystals. He embodied his results on naphthalene and naphthalene derivatives in his presidential address to the Physical Society in 1921, having been elected to the presidential chair in the previous year. He worked on the assumption that the benzene or naphthalene ring is an actual structure, which preserves its general form and size from compound to compound and, to the satisfaction of the organic chemist, his results justified this hypothesis. This work was the starting point of the series of investigations on different classes of organic compounds which he afterwards directed at the Royal Institution. These studies directly revealed for the first time the geometrical arrangement of the bonds between atoms, giving flesh to the structural formulae of the chemists and laying the foundations of stereochemistry. [An elderly lady, examining a model of naphthalene after a lecture at the Royal Institution, gave it a good sniff. 'It doesn't *smell* like it' she was heard to remark.]

After he came to the Royal Institution, WHB never did much experimenting with his own hands again, but it was a period of great team work. To quote, this time from W. T. Astbury, that enthusiastic member of the research team, 'Those were great days when he [WHB] re-made the Davy Faraday, and we all had our chosen tasks building with him – X-ray tubes and spectrometers, long-chain compounds, space-groups, basic beryllium acetate and other fascinating co-ordination compounds and organic crystals, X-ray photometry, the theory of rotation photographs, and all sorts of lovely things.' Having been asked by WHB one day to take an X-ray photograph of a fibre to show in a lecture, Astbury became interested in hair and other protein fibres. Shortly afterwards he went to Leeds as lecturer in textile physics, bringing the inspiration to textile research which WHB had seen as vital for the industry. (See the letter WHB wrote to Smithells on leaving Leeds, p. 139.)

J. M. Robertson, later Professor at Glasgow, was a member of the Laboratory in those days; he it was who worked out the crystal structures of anthracene and naphthalene.

Astbury again: 'Bragg himself was too busy with other duties to stick to any continuous line of research, but he lectured about all that was going on (and about much, besides, that was outside X-ray analysis), thereby revealing ever-new points of view, and from time to time he would dash back into the fray to thrash out something that had excited his eternally boyish interest.' The amount he published shows how actively he was engaged. After coming to the Royal Institution he wrote seventy-five scientific papers on his own, the last only shortly before his death. He also wrote a number of joint papers, as well as many popular articles and several books based on lectures.[3]

They seem to have been a cheerful lot working in the DF, they enjoyed themselves under WHB's headship; 'at the RI we were a happy family working together' George Shearer wrote to WHB after leaving the Davy Faraday. Astbury explains that WHB 'was not an active participant in everything, naturally, or the originator of every new idea, but he was the soul of everything. We who were his disciples and are his apostles had always the impulse to "tell Bragg about it" wherever we were working, for we knew that we could talk to him, that he would listen, and that he would be just as thrilled as we were. In the laboratory he would run to us just the same to tell us of any new idea that had occurred to him. He had a very human desire to share his joy, his boyish glee, and to be told that it *was* a good idea. His face would sometimes fall quite like a boy's if (very rarely!) you happened to think it was not a good idea after all – but you could always argue peaceably with him: he was your pupil just as much as you were his.' But here I would add that I can remember my father saying to me 'I cannot always understand when these young men come and tell me excitedly about their work; I have to say "leave your notes with me".' Then he would work through them and write out lucidly for the young man what the notes were really trying to explain.

The extremely important work relationship between father and son had been sorted out by the time WHB came to the RI. The outbreak of the First World War had stopped them working together suddenly and completely. When they got back to research after the war was over, it was natural that their work should tend to overlap since the interests of both stemmed from their pre-war collaboration.[4] WL at Manchester was wanting to build up a good department in crystallography; it was important to establish a field for his research laboratory. There is a letter from WL to his father showing that there were sometimes difficulties in sorting out the situation. They had had a discussion when WL got upset. He wrote:

> I am so sorry I was so stupid when you said you wanted to try rock-salt. It was awfully selfish and I want to take it all back. I don't feel like that for long and always realise things afterwards. It's just getting too excited over the research work that's the trouble. If only I thought I could help you like you help me with my work! . . . [though the child does not always want to be helped by the parent].

There was always rapprochement, and to make up for an outburst WL would write almost more remorsefully than he could continue to feel. There is evidence indeed that once the balance swung the other way, and in his autobiography WL records his subsequent regret:

> I wrote to my father with much excitement, and proposed that we should write a joint paper about it [the summing of the Fourier series]. He insisted that I should write it up alone, but I have always regretted that I agreed to do so. The idea as to how the Fourier series should be summed was entirely his, and I think he felt sad that the first paper describing its use should appear over my name. He should have published the idea, and I should have published its application to diopside.

They decided to divide the terrain.[5] WHB would attack the organic crystals; WL would work on the inorganic, metals, silicates, etc. 'Not that he [WHB] ever abandoned inorganic crystals or loved them any the less,' Astbury points out, 'witness, also about that time, what to my mind is one of the most beautiful things he ever did, the structure of ice; or his later return to quartz to explain so elegantly its modes of twinning.'

But in the main, father and son continued to research on parallel lines within their agreed domains for a long time; enough and plenty for each to explore grew out of the breakthrough of 1912–14. Together they put British X-ray crystallography in a position of pre-eminence which lasted for two generations. By the 'thirties WL and his school had brought order out of chaos in the silicate field and were making great inroads into the problems of metal alloys; while WHB and his pupils were demonstrating the power of X-ray analysis to solve problems in many branches of organic chemistry, such as aromatic compounds, carbohydrates, amino-acids, and so on. Astbury guesses that 'Bragg [WHB] saw, and saw more clearly with the passing years, that the crowning glory of structure analysis would one day be in the realms of biology', and claims for WHB, that, 'besides being by implication a crystallographer, he was by induction a biologist, a molecular biologist'. Lecturing in April 1928 at Guy's Hospital Medical School on the structure of an organic crystal, WHB stressed that 'interesting and significant as the new results are in relation to molecular structure, biologically they are literally of vital importance'.

Improvement of the research equipment in the Davy Faraday laboratory continued to be one of WHB's active interests. As late as 1936 two new X-ray generators were installed, a 'giant' X-ray generator, the pride of the laboratory, and a smaller one, each built to a new design by Müller. These generators accelerated experiments and made new advances possible.

As the keen researcher aged into the elder statesman (still full of youthful curiosity) a special faculty of WHB's came fully into play; an intuitive grasp of the implications of new work which he may not have been able to understand in its mathematical detail; but grasping the implications he could often give the work a new turn. Like Faraday, he sought and saw connections. Andrade wrote:

> In 1928, for instance, he arranged for Schrödinger to deliver a course of lectures at the Institution on wave mechanics, then a very recent development... He gave an introductory lecture as preparation for the course, in which he returned again to the wave and corpuscle dichotomy. He referred back with pleasure and some pride to his old experiments with Kleeman and Madsen [in Adelaide], saying, 'I may say, I think, that in these experiments we were, though unwittingly, carrying out Einstein's suggestion that the corpuscular hypothesis deserved careful exploration [see p. 74]... It is true, however, that I thought of the X-ray and γ-ray problems as distinct from that of light.' He ended on a typical thought, typically expressed, 'When the picture is finally clear there will no doubt be atoms in it,

> electrons, wave motions, energies, momenta and so on. But have we got them all rightly joined up? Perhaps wave motion belongs to more than the photon or to something else than the photon? We can only wait'. No doubt he was thinking of the experiments of Davisson and Germer and of G. P. Thomson, indicating the wave nature of the electron, of which accounts had just appeared.

Astbury noticed that WHB 'had the most amazing faculty of taking up a subject on which he had only the foggiest ideas to begin with, and quickly improving it out of all recognition; as, for example, at the Faraday Society's discussion on 'Liquid Crystals' held at the Royal Institution in 1933, when he explained the optics of smectic substances in terms of the cyclides of Dupin; and often, if you met him a few weeks afterwards, he would have reverted, not indeed to his initial fogginess, but to a state of comparative forgetfulness, just because he was pre-occupied doing the same sort of thing in some other line.'

But near the end of his life WHB's enthusiasm flamed up a last time for a piece of research in his old line; as Andrade explains, his interest

> . . .was aroused by the so-called extra reflexions or diffuse spots, which can be observed with powerful X-ray illumination of single crystals. These had been adventitiously observed from time to time, but in 1939 G. D. Preston published a careful study of them in certain simple cases and explained them as due to small crystalline fragments in imperfect alignment. The matter was immediately taken up at the Royal Institution, where the powerful tubes to which reference has already been made allowed the effects to be obtained with relatively short exposures. Mrs Lonsdale and H. Smith published a detailed study of the spots obtained with both organic and inorganic crystals, in which Bragg took the greatest interest. He was much intrigued by the effects, and was concerned about their explanation. The elaborate mathematical theories which had considered the thermal movements within the crystal as responsible for the reflexions did not appeal to him, and he evolved a simpler theory along the lines suggested by Preston, attributing the spots to a lack of regularity in the crystal structure. Dis-

continuities at the boundaries of the regular fragments of which the crystal is supposed to be composed lead to discontinuities in the phase relationship. Bragg wrote several short papers and notes on the subject, and, in fact, his last paper, written just before his death, and published posthumously, was on the secondary X-ray spectrum (extra reflexions) of sylvine.

It is this work that WHB was referring to in the letter written to his sister-in-law Lorna Todd in Adelaide, which is quoted on p. 121. But the subject had become a side stream; in 1939 war had diverted the main current of scientific research for the second time in the lives of my father and brother. Before the war ended, my father had died. His scientific heritage fell to WL. In the subsequent years WL guided and god-fathered the subject of X-ray crystallography which he and his father had started, until it was crowned with success by the triumphs of Perutz and Kendrew with protein molecules, Dorothy Hodgkin with vitamin B_{12}, and of Crick, Watson and Wilkins with DNA. Molecular biology, which had been started by Astbury, had now come of age.

Now that we have come to the end of WHB's years of research, it is fitting to recall the vision that inspired those years. Here is his own description – an extract from a lecture in which he quotes Michael Faraday's feelings which were so much his own:

> The story that we have to tell is great enough: in fact it is magnificent, awe-inspiring, as Faraday said. There is a passage in his diary which tells of his feelings whenever he began the search for a great truth at which he had dimly guessed:
>
> '25 Aug. 1849. I have been arranging certain experiments in reference to the notion that Gravity itself may be practically and directly related by experiment to the other powers of matter, and this morning proceeded to make them. It was almost with a feeling of awe that I went to work, for if the hope should prove well founded, how great and mighty and sublime in its hitherto unchangeable character is the force I am trying to deal with, and how

large may be the new domain of knowledge that may be opened up to the mind of man.'

In this particular case Faraday failed to find what he sought: yet the interest and sympathy with which we read his writing are not a whit the less. As a matter of fact, it was Einstein's brilliant reasoning, based on experiment of extreme difficulty, that in the end established the connection between gravity and electromagnetism in the existence of which Faraday confidently believed. All of us know something of that feeling of humility in the presence of greatness when we venture to ask a question of Nature herself. When we begin to assemble the instruments, quite simple they may be, with which we hope to enquire into that which is yet unknown, we feel a certain shame at our temerity and perhaps would begin our work alone.

That sense of the illimitable in its ordered magnificence which covers the student like a worshipper's garment, is a gift which no one may lightly decline. It is not only the researcher who can receive and appreciate the gift; it can be placed in the hands of everyone. It is the task of the researcher to describe what he observes so faithfully that his hearers also see the vision. He must serve them for their eyes, if they themselves are not trained to use them.[6]

Here is another passage in which WHB is talking of the research student and the task of putting his vision into practice in the daily adventure of research: and of how the spirit of research may inspire all men's lives:

A good research student is like a fire which needs but the match to start it. It is a discipline to put the text-book to one side and to get out further knowledge by one's own effort. It teaches the student how to value evidence; how to read with discretion, since he must weigh what others have done before he uses the previous knowledge as a foundation for his own advances. He learns to meet disappointment, to realise how little he can do in a day, and that weeks or months may go by without obvious progress. It is strange to discover that he must spend so much time on small things, that he must wait a week before he has succeeded in stopping a small leak, or go himself to buy some trivial thing, or spend weary hours in the adjustment of an instrument which in the end he learns to put straight in a moment. There is so much little work to be done before the good observations come; it may be that weeks are spent in preparation and five minutes in making the actual measurement. It is all very humiliating; and the blunders one makes are very foolish; and the one redeeming feature is that in its apparent perversity it is like every other piece of real work. Research is rather like playing against bogey at golf: Nature never has any weakness of which advantage may be taken; there is no hole to be won by bad play because our opponent plays worse. Yet research is very human, for the researcher finds himself one of a company who have in their turn striven and denied themselves, very happily; and have handed on their experience to those who take up the quest where they have left it.

It is, as I have said, a great discipline; and there is always the hope before every student that he may contribute something to the total of human knowledge. Perhaps the hope is not so often reached as it might be, but that does not mean that the work has been done in vain: a thousand times, no. It is a remarkable fact that if a man tries to set down on paper, in line or in colour, that which he sees in Nature, his own vision of what is beautiful is quickened. The pencil teaches him the beauty of line, and the brush opens his eyes to all sorts of delicate colour schemes which he never saw before. It does not matter if his efforts are a bitter disappointment to himself, and he need not show them to any one else. Similarly, the man who strives to understand the workings of Nature by experiment which may make him feel very feeble and stupid, is paid by his discovery of a richer world. There is a fellowship between all who have tried to understand, which enables the worker in one field to have a welcome power of appreciation of the work of others. He gains not only in the richness of appreciation of what is beautiful and interesting, but also in the power of making friends. . .

Here are various reasons for the encouragement of research: the benefit of the student, the addition to human knowledge, power and riches, and the needs of defence, military and industrial. But I think we still have failed to include the most important reason of all, the real reason of which the others are only derivatives. It is that the spirit of research is like the movement of running water, and the absence of it like the stagnation of a pool. Scientific research, in its widest sense, implies, of course, far more than exploring the questions of physics and chemistry and biology. It is not a religion; but it is the act of one. It is the outcome of a belief that in all things which we try to do we may by careful seeking and by a better understanding do them better; that the world, far beyond what we can see of it on the surface, is full of things which it would be well for us to know. It is our duty and our gain to explore: we have always grown by doing so, and we believe that the health of our souls depends on doing so. Shall we sit still when there are difficult questions to solve; and when the answers may give us new insight and new power? There is a hesitation which would beg us not to push forward lest we come to think less of the world. As against this, research is an act of faith in the immensity of things. There is no end to the search: it is a poor thought that there might be.

The spirit of research would drive us all to work to the utmost of our power, believing that the more we do and the better we do it, the better for the work and lives of others. It is vigorous, hopeful, trustful and friendly; it adds always new interest and new life. It is a spirit which should run through all our activities, and not be found in laboratories only. It is, in fact, a spirit which is essential to us as a nation trying to rise above ourselves to better things. All our efforts to encourage research have before them not only one or other such immediate object like that to which I have already referred, but also this great ideal.[7]

The young professor at Adelaide

WHB and his spectrometer at University College London

GB in early years at the Royal Institution

Father and son at the British Association, Toronto 1924

WHB in Faraday's (and his own) study at the Royal Institution. By lighting a candle (in Faraday's candlestick) he turned on rows of lights in the Museum of Science and Technology at the Rockefeller Centre, New York, during the opening on 11 February 1936

He was a splendid grandfather; Watlands 1940

10

EDUCATION, INDUSTRY AND NATIONAL AFFAIRS

In this chapter I want to tell how WHB worked for education, and for industry and for national affairs; in the end becoming something of a national figure representing science. The urge underlying this work was his deep sense (one could say almost a religious sense) of responsibility.

We must go back a long way. It all began in Adelaide, and with education. I have said how disappointed WHB was with the quality of the students he found in his laboratory when he arrived. At the University Commemoration in 1888 he gave an address on the state of education in Australia and at home. He was only 26 years old; he was very shy; he had only arrived in the colony two years before. But he had been so shocked at the state of things that he boldly prepared and gave this address. The *South Australian Advertiser* for 2 December reported it at length (and if this work on education in Australia seems a bit dull, remember how important it was to the earnest young professor).

The gist of the report was this. In the last decades of the nineteenth century technical education had advanced on the continent of Europe far further than in England. There was considerable fright, as the leader writer said, that 'the competition of foreigners, who have received a commercial training, is taking the bread out of the mouths of English clerks whose education has been of a general rather than of a special character'. There was a clamour, too, that 'working class' children should be taught trades in their school hours. WHB in his address spoke out strongly for general education rather than vocational, saying it was not so much the subjects taught but the method of teaching that was at fault. The secondary schools in Australia were modelled on the English Public School pattern with their almost exclusive emphasis on ancient languages and mathematics, with a little text-book science. He pleaded for experimental science in the curriculum, pointing out that training in observation and logical deduction would be of service in whatever career pupils followed later. WHB went further, to ask whether this experimental method might not be employed in other subjects as well, it being the natural way in a child's progression to start with the concrete and go on to the abstract. The classical system of education

> may develop in the young generation the capability of fulfilling duties in certain traditional ways but it does not so train minds that, having a knowledge of the tools that modern science provides, and judgment as to what may be done with them, they may strike out for themselves new kinds of work and new methods of working.

(WHB always grudged the time and agony he had himself spent on Latin verse. Of Greek thought and wisdom he was given nothing.) He ended his address, not very hopefully, by saying that only an informed public opinion could ever bring about a change.

It was not a new point of view. Humphry Davy had advocated science even as a training for young ladies. He opened his 1811 series of lectures in Dublin with the words: 'In this room I am sure that I need not enter into any elaborate arguments in favour of a certain acquaintance with the philosophy of nature, in the system of improvement of the female mind.'[1] Faraday, in 1862, declared in his statement before the Public School Commissioners: 'That the natural knowledge which has been given to the world in such abundance during the last fifty years should remain untouched and that no sufficient attempt should be made to convey it to the young mind growing up and obtaining its first views of these things is to me a matter so strange that I find it difficult to understand; though I think I see the opposition breaking away, it is yet a very hard one to overcome. That it ought to be overcome I have not the least doubt in the world.'[2]

About ten years after his first address, Professor Bragg's views were again given newspaper prominence. Again there was a first leader (in the *South Australian Register* 9 May 1899) headed 'Schools & Teachers', followed on the next page by 'Education in England and on the Continent. Report by Professor Bragg'.

This report was the harvest of a year's research in England.[3] It is evidence of the attention paid to WHB's urgent enthusiasm that, when he went back home on sabbatical leave in 1897, he was commissioned to collect any information about education in England and on the Continent that would help solve South Australia's educational problems.

While in England he travelled around; we know he visited schools in Birmingham and Cambridge (convenient because GB and the boys could stay with GB's married sister); he inspected the Battersea Polytechnic (and was intrigued by the new idea of schools of Domestic Economy): he interviewed and corresponded with leading educational authorities, and collected details respecting schools and school systems in England and Wales and in other countries including France and Germany (especially Prussia) and Switzerland. He attended a Teachers' Guild Session at Aberystwyth (combining it with a bicycle tour with GB through Wales).

On his return to South Australia WHB presented his report to the Minister for Education (the Hon. R. Butler) which the *Register* leader-writer described as 'necessarily somewhat lengthy'.[4] A summary is given in three columns, and this is a summary of the summary.

> The key of the [educational] situation in England [WHB wrote], is that English people conceive themselves to have been outstripped by their continental neighbours. . .and that they are making vigorous, though somewhat confused efforts to retrieve their position. [They were as worried in 1899 as they had been in 1888.] Certain questions that arise out of this situation are eagerly and widely discussed. Two of the most important are:
>
> the provision under State authority of education other and higher than primary;
>
> improvement in the methods of providing and training teachers.

He went on to praise the sound and ordered system of education on the continent, mentioning Saxony where education had already been brought within the reach of all.

> The ethical results obtained in [England's] Public Schools are unmatched on the Continent; but in the nature and quality of much of the instruction given there, in the provision. . .of schools adapted to the wants of various portions of the population, in the systematic connection between the various kinds of schools. . .[England] has much to learn.

He hammered the old point:

> Good general instruction ought in all cases to precede any

> attempt at good technical instruction. . . There can be no doubt that, from want of it, much of the money spent on technical education in England has been simply thrown away.

And then, about hopes for teacher training reform in England; the Pupil Teachers' Departmental Committee of 1898 had said 'We think it extremely desirable that all intending teachers should pass through a secondary school for the completion of their education.' The Adelaide leader-writer pointed out how hard was the lot of the pupil teachers who must teach all day and study all evening to pass their qualifying examination.

The result of all this was that in 1900 South Australia made compulsory a year's training at the University for all elementary school teachers; not only compulsory but free, owing to a generous benefaction to the University. The Governor, in a speech made in 1903, claimed that in this South Australia had led the world.[5]

But however much WHB was against vocational teaching in school, he worked hard for technical education in its right time and place. He was much concerned with the School of Mines, and with the regulations governing elementary commercial examinations; as he said:

> the panic stricken haste which had attended the setting up of Technical Schools in England. . .emphasizes the need of a firm basis of general knowledge.

When the Royal Commission on Melbourne University visited Adelaide University in 1903, they declared: 'We are greatly impressed. . . One prominent characteristic is the liberal attitude the staff display towards education generally'.[6]

Then in the next year, 1904, came WHB's thrilling adventure into scientific research, and there are no more press cuttings about education until he is about to leave Australia in 1909, when, among other farewells, there is a touching good-bye from the teachers (he had set up a branch of the British Teachers' Guild); there is a cutting about WHB advocating a Chair in Astronomy at the University, which must have pleased his old father-in-law Sir Charles Todd, the retired Government Astronomer with whom he had spent happy times experimenting with wireless telegraphy; and then there are full accounts of WHB's Presidential Address to the Australasian Association for the Advancement of Science at Brisbane in January 1909, his last public words to Australia before he sailed.

The first part of the address was an account of the latest developments in radioactivity; the second part a review of what Australian universities might do, with an urgent plea for research scholarships to keep graduates at the Universities for a period to do research in pure and applied science. He could say, by 1909,

> In most of the States there is a more or less effective educational ladder from the primary school to the university which is much used already, . . .[and] the training which is given in the university course is well suited for students who may afterwards take up research work under proper direction. . .[but] most of them must at once set to work to earn bread and butter and they scatter far and wide.[7]

He pleaded for co-operation between University and the needs of industry and agriculture:

> Pure science and technical science draw life from each other, and must on no account be separated.[8]

There was need for research into new methods of mining, in wheat breeding, viticulture, sheep disease, orchard disease and forestry.

> If the State University is to live its full life it must not separate itself into the wilderness like the hermit of old; but must mingle with the people and draw strength and inspiration from the attempt to minister to their needs.[9]

The reporter of Brisbane's *Daily Mail*, who obviously had not understood the first part of the Address on Radioactivity, was cheered by the second, and wrote: 'When Professor Bragg directs attention to the spending of money and time on research work and the effect it is likely to have upon our mining, manufacturing, and agricultural industries, we can all enter with zest into the discussion'.[10]

WHB was in his 47th year when he arrived at Leeds in the spring of 1909, having completed nearly half of his career, and that mostly given to the cause of education and teaching. It was a long training in the art of exposition. Though the confidence gained at Adelaide was nipped at Leeds, the training bore fruit later in his Royal Institution lectures (especially those for children), in addresses, broadcasts, and in his speeches as President of the Royal Society. Once a rather brash young man told WHB that he was going to be a writer; WHB looked up at him saying quietly 'Oh, what about?' Words were tools for a purpose, to WHB: his purpose, not only to tell of scientific discovery, but to tell the ordinary man what science was about, and to awake wonder in young minds. WHB cared for tools, and his words have the grace of fitness in conveying his thought.

During the years as Professor in industrial Leeds WHB got to know manufacturers, and lectured to the Workers' Educational Association on top of a heavy University lecture load. Research, which at first failed so agonisingly to prosper, blossomed brilliantly after 1912 in collaboration with WL, and in 1915 came the invitation to a Chair at University College London. WHB accepted and wrote the letter to Professor Smithells (Professor of Chemistry, senior and influential member of the University Council), to which I have already referred on p. 80.

The letter explains his resignation from Leeds: it is long and here are extracts. They show WHB's first interest in education broadening from school, technical school and University into industrial affairs; and show him exploring the possibilities of the scientist's contribution to national affairs.

The University,
Leeds.
March 26 1915.

My dear Smithells,

...I think if I go to London the research work *will* get on faster... But as I look at it, I think I see that it is not merely a question...of research facilities...

After the war, if all goes right, we know, as you say implicitly, that there is another struggle coming on, in which the organisation and efficiency and well-being of England must be considered from the point of view of what science can do to improve them. There will be many great movements required: it is up to all of us to help, each in the one he can do most.

Though the introduction of the scientific spirit and method into manufacture is immediately applicable here, yet it is a part of a general movement of the whole country, and cannot be considered by itself.

The physical laboratory in Leeds should develope itself into a research laboratory for the industries of the Riding. Its professor and staff should have before them, as the ideal behind their work, the investigation of all physical questions that are involved in the textile trades to begin with. They are worthy of a man's whole time: several men in fact. . . In this way you start a true physical research laboratory for the district. You will not get your manufacturers here to approach it the right way without an example, they are too ignorant, stupid and jealous, with honorable exceptions. I quite understand, as you know I do, the tale of qualities on the other side which have made Yorkshire the great industrial centre that it is.

You see, the Textile department does not know enough physics: it could not be expected to. But the physics problems are not to be solved except by good men, they must be taken very seriously.

I could make a shot at it myself; but I am not so well equipped as many younger men and I should have to give up all my own research work, which would seem a wrong thing to do, as far as I can judge.

I don't think I have been useless in this direction the last few years because the people do respect physics more than when I came, I think: . . . But now is the time for action; and if we could find a keen young man to take my place and to throw himself into the work which I have outlined, that is I think the next stage.[11] Let him prove to them against their own convictions that scientific examination of their problems is the right thing: and chuck important solutions at their heads. The war will help him.

On the other hand another part of the *same* work can be done in London: and I think that is where I might hope to come in. The Government will want help: I believe they are sure to do a lot, and in time I might get a footing in their councils. Moreover there is the Royal [Society]. Could not we make it a real advisory body, with more worthy work than that of reading papers? I know there are several already doing their best, but I think I might help: I have

> already had personal hints to that effect. The Royal ought to have its eye on all the possibilities of encouraging scientific application, should know what is to be done, be in touch with Universities and research laboratories, be ready to advise and co-ordinate and to acknowledge good service. I must not undervalue what it does already: but I think it could help so much in what is coming.
>
> I think I could help all the more because I have been here, and lived in the centre of these industries, and got to know something of the character of the men & the masters, their wonderful good qualities and their ignorance of certain aspects of their work.
>
> It seems to me all one thing, and that I may be serving you and the general cause better by going now.

He went, late in 1915, but did not stay long in London. He was soon away on war work for the Board of Inventions and Research.

Before he went north to be Resident Director of the Admiralty Research Station on the Firth of Forth he was interviewed: there is an article in *The Saturday Evening Post Magazine*, New York, 4 March 1916, 'British Science and the War. Interview with Professor Bragg at University College'.

The first part of the article is about the need for a permanent research establishment for the Navy; then about the desperate need for munitions, and also optical glass and dyes which used to be imported from Germany, which difficulties lead on to 'the more general question of the attitude of English manufacturers to science. The typical business man has been so severely blamed of late for the narrowness of his outlook' states the reporter: but he found WHB refusing to join in the chorus of censure. WHB confessed that when he came home from Australia a few years before he was inclined to think the English business man a laggard, but went on:

> It is quite true that it is unusual for the head of a big English factory to employ a staff of trained chemists on research work, in the hope that they may discover some process that will be useful for his trade, but you must remember that the British manufacturer already has his hands full. He has as much as he can do to meet the demands for the products he is accustomed to make.

WHB pointed out that it is in the countries fighting to get their goods accepted and sold that manufacturers encourage research. However, 'Such contrast in practice does not mean. . .that British business men are indifferent to science' and here WHB instanced what was being done to try to bring science and industry into closer co-operation at Leeds and the newer universities in the Midlands and the North.

A day or two before, a memorandum had appeared in the press which obviously had occasioned the reporter's visit to University College: it was 'a manifesto by several prominent British men of science who are dissatisfied with the general attitude of British Governments to scientific matters. They call attention to the lack of scientific knowledge on the part of public leaders and administrators. . .and point out that in the whole history of British Governments there has been only one Cabinet Minister – the late Lord Playfair – who was a trained professional man of science.' The reporter continued: 'I found Professor Bragg prepared to go a long way in endorsement of this memorandum. He urges that there should be a Minister of Science, not for the duration of the war only, who should be responsible for the national co-ordination of research. It would be the duty of this Minister to acquaint himself with the resources of the British scientific world alike in personnel and equipment. . .he could have warned of the dangerous position about glass and dyes and could have prevented the export of lard from which Germany had got glycerine for explosives.' Meantime something could be done by the Royal Society. 'Until a Minister of Science is appointed Professor Bragg would like to see the Royal Society undertaking something of that function of the co-ordination of research which should ultimately be the province of a member of the Government. . . Another proposed change for which some British men of science are eagerly pressing just now, does not commend itself to Professor Bragg's judgment. They are loudly. . .advocating an entire recasting of the educational system from the foundation, with the idea of making scientific subjects the main feature of the curriculum of every school. Professor Bragg does not share either their contempt for the educational value of the 'humanities' or their belief in the complete sufficiency of a scientific course. . . It is to the due combination of these two elements that Professor Bragg looks for the ideal education.'

So it all comes back to education again. But out of all this unease the Department of Scientific and Industrial Research was born.

In July 1915 a White Paper had been published which found that there was a strong consensus of opinion among scientists and industrialists that a need existed for new machinery and State aid to promote industrial research. Only a week later an Order in Council set up a Council for Scientific and Industrial Research to be administered by the Board of Education. In December 1916 responsibility was transferred to a newly created department, the Department of Scientific and Industrial Research.

The Privy Council reserved £1m. to finance research, and especially to provide grants for co-operative industrial research organisations.

WHB had thought that the Royal Society might organise this work, but it could not have provided the funds. The advisory Council of the DSIR, however, was composed of Fellows of the Royal Society, and in his Presidential Address to the Royal Society years later, in 1939, WHB spoke of the DSIR 'which we are proud to think of as an original device of the Society'.

The 1914–18 war first woke some public understanding of what science could do for industry and defence: by now, after two world wars, it is taken as a matter of course. To quote from a history of the DSIR: 'The "running-in" period proved indeed to be one of great difficulty and anxiety, and at times there were those who were inclined to write-off the whole experiment, from the Government point of view, as a failure.' But somehow the Council managed to convince successive Governments that it was worth carrying on. 'Up to 1934, although some [research] associations were well established, the future of the movement as a whole remained precarious'; . . .however, 'by the beginning of the 1939–45 war, it seemed that the corner had been turned'.[12]

But the response of the captains of industry had been grudging and slow. A hundred years before they would not have been so suspicious. In the 19th century Faraday did much for industry, and his help was thankfully accepted by men who could not understand but wanted to use his electricity. These masters who welcomed the new inventions appreciated the advantage of new methods, for they were close to the workers in their small units, had probably worked with their own hands and seen how science could multiply the work one pair of hands could do. But the sons and grandsons of these men, educated on money made by the machines, began to take form as a new class in the nation, and distance increased between them and the job and the workers on the floor: and the ranks of the new managerial class were swollen by the sons of the older professions and the gentry who began to go into 'business' as they would have scorned to do half a century before, because now they saw there was money in it.

The new 'business' gentleman was taught little or no science at his Public School; there was practically no teaching of experimental science when WHB was a boy, in the 'seventies. And when science did begin to creep into the curriculum, 'I can remember how science teaching was looked upon askance by the other modes of teaching' wrote WHB; the science masters were the lowest grade in the Common Room. Yet while the upper and middle classes remained ignorant of science, the clever boys of lower class were beginning to snatch the chance of rising in the new subject, boys whose background made their excelling in the old disciplines unlikely or impossible. Some years ago now, my brother made an analysis of the Fellows of the Royal Society which showed that only one in fifty had been at Public School. In the nineteen-twenties and nineteen-thirties an educated voice could exclaim 'Of course I know nothing of science' as though it were an amiable weakness to be applauded: that at the one extreme, and at the other a bright young scientist could be narrow-minded, slightly contemptuous of all but science, and lacking skill in words to make his findings plain. It is little wonder that there was a gulf and that science found it hard to get a hearing.

From 1928 to 1931 WHB was chairman of the DSIR committee on the application of X-rays to industrial research. Especially during that period, but also before, and later as President of the Royal Society, he went about visiting works, opening research laboratories, lecturing, broadcasting, speaking at luncheons and dinners to try and explain and encourage the scientific outlook. I remember his frustration at the slowness and obscurantism in some industries; but he plodded on, speaking and writing on 'Science and Industry', 'X-rays to the help of British Industry' and so on. Press cuttings (carefully collected by WHB's secretary Winifred Deighton and pasted into scrap books in slightly haphazard order by her 'help' Violet) give account of what he said to the gas industry, optical people, dyers, to the gemmologists, to the foodstuffs, electrical and kinematography industries; to engineering and transport workers. He likened the new metallurgical laboratory at Sheffield to the restraining constituent in an alloy '...it holds theory and practice together and gives the industry strength and tenacity': he opened the launderers' research establishment, spoke to the rubber people, to Pilkington's glass and so many more, and I remember going with him when he went to talk to the boot trade at Kettering and we stayed the night in kind and plushy comfort and were each given a pair of shoes.

As he struggled to get industry to take advantage of the new knowledge, WHB absorbed the old lore. It fascinated him to discover the explanation of old ways of working. One outcome of this was his set of Children's Lectures given at the Royal Institution at Christmas 1926, 'Old Trades and New Knowledge': another, his Presidential Address to the British Association at Glasgow in 1928 on 'Craftsmanship and Science':

> The growth of knowledge [he said] never makes an old craft seem poor and negligible. On the contrary it often happens that under new light it grows in our interest and respect. Science lives on experiment; and if a tool or a process has gradually taken shape from the experience of centuries, science seizes on the results as those of an experiment of special value. She is not so foolish as to throw away that in which the slowly gathered wisdom of ages is stored. In this she is a conservative of conservatives.

In 1935 he returned to this subject, speaking at University College, London:

> It is a fascinating business to take some old precept handed down by tradition and to ponder over its origin and its meaning...
>
> Our country folk talk of the 'blackthorn winter' or the days of the 'icy saints'. They are convinced that the advancing spring is liable to a sharp and sudden setback when the blackthorn is in flower, as, indeed, happened most disastrously in this present year. Surely their observation is not at fault, though their description is quite irrelevant. An enormous mass of cold air moves down from the polar

regions to the equator in spring. It is not surprising that in its turbulence a mass of it may descend upon us for a brief period, at a time which is not very definite, but is apt to coincide with well-marked events of the year...

It has long been known that the barberry plant has an evil influence on the wheat crop. In Massachusetts...the barberry was once declared a public enemy, and a law was passed ordering that it should be uprooted. The plant biologist can now give us the reason. Rust is a deadly fungus, a perpetual plague... The variety that is responsible for damage to wheat has a two-year period of incubation, and one of those years is spent with the barberry as its host. I suppose that the discovery of this habit of the rust fungus was based on a knowledge of the old tradition...

Why does the smith as he plunges his red hot iron into cold water stir vigorously at the same time? Presumably because the water next the iron must be changed frequently so as always to be cold, perhaps that the temperature at different parts of the surface of the iron must be the same. The iron worker uses an iron hammer to drive an iron chisel when he wants to chip an iron mass. But the carpenter uses a wooden mallet and a chisel with a wooden handle. In the former case a very sharp blow is required to shatter the hard crystallites which we now know to make up the body of the iron. In the latter case the blow must be longer and more uniform, more of the nature of a shove. The stone-mason's work lies in between; he uses an iron chisel and a wooden mallet. The metallurgist has evolved in the course of the ages a number of curious receipts for the furtherance of his work. He has found out the effect of the addition of mere traces of foreign substances to his materials: such as the easy flow of steel from the ladle when a little aluminium has been added. But many of his devices were purely empirical; only now do we begin to understand the fundamental causes. The maker of stained glass discovered by accident that the inclusion of finely divided gold into his melt gave him a glorious red: but he did not know why. The spinner discovered that a certain amount of moisture must be present in the air if his work was to be successful. He must bring up the humidity of his mills to a certain standard, and

> it was better that he should work in Yorkshire or Lancashire where that standard was more easily attained. He did not know, of course, that his fibres became electrified by stretching and rubbing and repelled one another. The fine subsidiary fibres which project from the main fibre could no longer interlock and link the main fibres together.
>
> It is an old observation, as old as the first days of the pitching of tents, that a rope is shortened by wetting. This is curious, because hair fibres are not shortened thereby. But a closer examination of the behaviour of hair and wool shows that moisture also causes fibres to swell sideways. The rope grows thicker as it is wetted, and the fibres that are wound about it in spiral fashion cannot encircle it so many times. Thus the rope as a whole is shortened.
>
> Thousands of other instances might be given of old precepts, wise sayings, and able craftsmanship which are worthy of record and of examination in the light of modern knowledge before they are forgotten.[13]

As WL used to say, the job of getting science used had been a job rather like that of an underpowered locomotive getting a train started uphill; but by the late nineteen-thirties momentum had been gained and the train was speeding.

> The work of discovery goes on and no one can stop it, not even ourselves [said WHB, lecturing in New York early in 1939[14]]. The constant demand for knowledge that is required for the solution of problems in health, in industry, in every human activity is so insistent that knowledge increases continuously and rapidly. And even if there were not this practical urge there would be the never-failing curiosity to know more. We must therefore accept the position; we all seek for an understanding of how to make the best of it.
>
> [There were a few obvious lines of action, among them] there is the great question of right exposition. It may be that there are some who would even now disclaim any duty of scientific men in this respect; and certainly there were many who would have done so in the past. If, however, we suppose that natural knowledge and the power which it gives are a common possession of mankind, we ought to make

> sure that what is found is understood. We cannot compel men to make use of science in the right way, but the chance that good use will be made is in a curious way dependent on the ease with which it is stated.

And so, back to Education. WHB, speaking at the English Association dinner, said:

> And now I come to a point where I would gladly take counsel with you. I have become convinced that the means of communication between those who explore the new fields and the great majority who necessarily stand outside, are far from being as efficient as they might be, and that an improvement is overdue. The defect must lie in the language which is used for the interchange of ideas. . . It may be justly said that it is impossible to speak the truth to a number of people at the same time. Now when the matters to be described are unfamiliar, and language has not been devised to describe them. . .the chances of misapprehension . . .are greatly increased. So the scientific worker finds it difficult to convey the interest, or importance, or beauty of his discoveries. . . Skill in the use of language. . .in the choice of words. . . This is surely a fine use of language and worth teaching.[15]

And he amplified this on another occasion:

> The story which science has to tell requires a statement which must especially be simple and clear, which transfers knowledge and inference from the mind of the writer to that of the reader with as little misunderstanding as possible, and these are some of the qualities of literature at its best. The science story is *sui generis*, but the good telling of it is no less difficult than that of any other.[16]

Always an educator; wherever he went WHB pursued his crusade. He was President of the Science Masters' Association, he opened school laboratories and spoke at prize-givings; told Roedean that 'the humanities and science were not rivals but companions and neither could grow without the other'; and to the Royal Society insisted that science and the other 'humanities'

> have far more in common than is generally recognised. [And he continued] men of different types of mind must be kept

> together more than at present... It happens too often now that the administrator makes mistakes because he cannot understand his technician's advice, and the technician is incapable of expressing himself so that his administrator understands him. There ought not, in fact, to be a sharp distinction between the two.[17]

He told St Edward's School Oxford that 'the type of man needed is a scientific administrator'; and explained in his 1936 Presidential Address to the Royal Society how research workers in industry could come to a blind alley, saying that if their education were improved and made more general they could then move into the councils of industry,

> places of responsibility to which specialists in science are as yet rarely admitted... The scientific expert must himself help to take down the barricade that makes the alley blind. This requires that his education should be much more than sufficient to make him only a laboratory man: which brings us back again to the very important point, that the man himself must be as much the care of those who give him research work to do, as is the work which they set before him.

Foreign languages also came within WHB's attention.

> You do not need to know another man's language so much if you are going to fight him, [he told a school assembly] but you do want to know something of his language if you are going to help each other.[18]

Help your neighbour, help your country, explore and experiment, communicate...

In his 1938 Presidential Address, WHB spoke of scientific responsibility.

> It is curious to think that the feeling of responsibility is of recent growth. There was even a time when a discovery could be considered a private possession, to be withheld from general knowledge if thought fit...Rumford, at the end of the eighteenth century, was one of the first to accept and proclaim his responsibility, and act upon it, when he tried to apply the laws of heat to the economy of fuel... The Royal Institution, in its first form, was his attempt to give concrete expression to his idea. The founders of the British Association were urged in their work by a similar motive

> and were to some extent under the belief that they were repairing a defect of the Royal Society [see an interesting letter about the British Association from WHB to Sir Oliver Lodge given in full at B in the Appendix]. . . The Royal Society undertakes a special part of the general responsibility. It is concerned with affording opportunities for the publication and discussion of original discovery. . . [But later he complained that] the volumes of [Royal Society] *Proceedings* and *Transactions* have a certain resemblance to a building site on which the contributors of materials have shot their goods. . .without regard to convenience of subsequent handling. [And he welcomed the recent plan for summaries which recognised the importance of exposition by making the essence of papers available to specialists in other lines of research. The Society] also encourages discovery by the administration of funds entrusted to it for that purpose. [By this time large benefactions were coming in due to the success of science in industrial, medical and agricultural research]. And of course, it represents officially the scientific activities of the nation, acting as an adviser on such matters in public affairs. In this field it has full opportunity for the exercise of all its powers

but only as consultant, and government was not always anxious to make use of scientific advice.

The Government had woken up to the usefulness of science in war during 1914–18. The success of the DSIR, slow as it was to start with, made its impression and the work of the National Physical Laboratory had grown in strength. It is evidence of this that in 1934 the Prime Minister presided at the first of a set of lectures organised by the British Science Guild: 'To keep those who run national affairs in touch with the latest developments of scientific progress.'[19] WHB gave the first lecture, Rutherford the second.[20]

All down the years WHB had been hoping to get the voice of science heard in state counsels – one reason why he had worked so hard to train and educate the voice of science and also to persuade science, reluctant to leave the laboratory, that she must accept responsibility in outside affairs.

Early in 1939 the Royal Society presented to the government their carefully compiled National Register of Scientists containing some eight thousand names. (Cockcroft had suggested making this list and WHB had supported its compilation before ever Sir John Anderson began to place the National Service organisation on an efficient basis. The seed for the idea of making this list had probably been sown by Hankey's and Tizard's discussions in the spring of 1938 on the using of scientists in the forces.)[21] The Register was not received with enthusiasm by the Chiefs of Staff. At an inter-service meeting early in the war, WL was told that research was a peace-time occupation; he must realise that now that war had broken out the armed forces had more pressing duties and were unlikely to be able to employ more than fifty scientists. But within a year hundreds were being asked for.

In reviewing the situation of science and the war in his 1939 Presidential Address to the Royal Society, WHB made four statements, which summarised are:

1) Science is of fundamental importance to the successful prosecution of any enterprise.
2) Science is of general application; there is not one science of chemistry and one of medicine, not one of peace and one of war. There is only one natural world and one knowledge of it.
3) Fruitful inventions are always due to a combination of knowledge and of experience on the spot. Unless the man with knowledge is present at the place and the time when some experience reveals the problem to be solved he misses the fertilising suggestion. Neither can the mastering idea suggest itself to the man who has the experience only but no knowledge by which to read the lesson that the experience teaches.[22]
4) There are difficulties peculiar to the application of science to war purposes due to the weak connection between pure science and the Defence research, which must be secretive.

> The tremendous use of science in the present war compels us to think about the method of its use. Many suggestions have already been made: as, for example, that a Ministry of Science should be formed immediately. In my opinion there is no solution here. A ministry would be too formal and rigid . . .the most successful ways of using knowledge are personal and elastic. We must not attempt too much at once.[23] We might be content if we could in some way bring Science

> as a whole into close relation with Government as a whole, if we could attach a central authority of science to the central authority of the country. The immediate application of science in any department of the country's business should be made from within the department, not from without, and we have already a number of instances in which this principle is followed. . .[such as the DSIR, the Medical and Agricultural Research Councils]. Each department of the Defence forces takes care of its own applications of science. This is satisfactory so far as it goes. The need is rather for means whereby the Government, in its care for the whole sweep of the country's business, can rely on and make use of the whole range of scientific knowledge.
>
> It would seem that our Society itself is a body which is not used as it might be.

And he went on to make some suggestions as to how the Royal Society could better be used.

However, WHB, as President of the Royal Society from 1935, was made chairman of a committee on the potentialities of home produced substitutes for imported fuels; their report was made within the month: and in June 1940 he was made chairman of a committee to advise on the scientific aspects of the Government's food policy with terms of reference 'to consider and advise upon problems of national food requirements and of home food production, with special regard to the shipping and foreign exchange likely to be available for imports of food and animal feeding stuffs, and the labour and other resources likely to be available for home production'.[24]

WHB organised a set of lectures at the Royal Institution, 'to spread more widely a working knowledge of the fundamental principles of nutrition'; they were published in book form as *The Nation's Larder and the Housewife's part therein.*[25]

WHB was in request as a chairman, and not just as President of the Royal Society, because, as he almost boasted, but very modestly: 'Committees seem to get along together when I chair them'. He could draw a committee together so that it 'found itself' like Kipling's ship, and got somewhere. Perhaps that was why he was roped in as chairman of the Anglo-French Ambulance Corps. Various English towns and associations were to send ambulances to the French – a worthy object, but hardly WHB's subject.

The Presidential Address in which WHB had voiced his hopes and anxieties for the use of science in the war was given in November 1939; by June 1940 the Royal Society was making concrete suggestions to the Government. WHB wrote, as President:

The Royal Society,
10 June 1940

Dear Prime Minister,

The appointment by the Lord Privy Seal of the Committee to advise on Food Requirements and Production, of which I have been invited to act as Chairman, affords clear evidence of the Government's determination to utilise all the resources of science in solving the problems of the war. [This is a very tactful beginning, *but*:] if these resources are to be applied quickly and effectively. . .it is my own view, and that of many of my colleagues, that there is urgent need of a piece of co-ordinating machinery which does not at present exist. . . The need for such co-ordination, so far as scientific research is concerned, was fully recognized by the Research Co-ordination Sub-Committee which reported in 1928; but the machinery then set up has since largely lapsed, and a method which may have been adequate to the needs of peace-time is quite inadequate in time of war.

WHB explains in measured sentences, and sets out the Royal Society's ideas for the composition and aims for this committee which should be 'a focus point. . .at present lacking in the organisation of our national scientific effort'. It is a long letter, quoted in full in the Appendix section D.

I found this letter at the Record Office. With help and some difficulty I had dug out a big file of correspondence with the Royal Society[26] from among the Cabinet Papers, newly released in 1972. It was moving to encounter my father speaking from the first page.

Following WHB's letter are minutes flitting to and from secretaries asking what was happening; then a long letter from Sir Alan Barlow (a Treasury Secretary) inviting WHB and officials of the Royal Society to meet him. After their discussion WHB wrote this:

> The Royal Institution
> 24 July 1940
>
> Dear Barlow,
>
> At the interview which you kindly gave to Egerton, Topley and myself a fortnight ago, you asked for examples of cases in which the committee proposed by us would have been of use to the Government. I have been thinking much about your enquiry.
>
> If we try to give details of particular instances we come up against the difficulty that to be effective in this way we must accuse individuals of neglect and inefficiency. It would be a horrible task to get up a case involving the preparation of such evidence...

WHB could not have done it. Instead he gave an example of faulty strategy – of how lack of chemical engineers to run the huge armament factories in the 1914–18 war had led to a new department being created to train them and then through lack of co-ordination, advantage was not taken of that expertise in the Ordnance factories that carried on between the Wars. The letter contains some shrewd comments and ends with a fresh setting out of the Royal Society's ideas. The full text is printed in the Appendix section E.

One can imagine Sir Alan Barlow then settling down to make out his own 'Notes on the Royal Society's Proposals' from the Government point of view. After outlining them he commented, 'there is a general malaise among scientific workers, especially the younger men, who feel that the scientific personnel of the country is not fully utilised, and that the Ministers and Departments tend to ignore the scientific aspects of the problems of Government'.

Meanwhile Professor A. V. Hill was writing (10 September 1940) from the Royal Society to Group Captain Elliott (of the Cabinet Secretariat) enclosing a memorandum he had compiled on 'Scientific Research and Technical Development in Government Departments' which he proposed sending to the Lord President of the Council (Neville Chamberlain), saying to Elliott:

> Egerton strongly feels that something must be done about the business, or there may be an explosion: anyhow I propose

to make myself a nuisance until something is done or I am squashed.

More letters, more notes; and then in the file is a Minute from Sir Alan Barlow to Neville Chamberlain:

Treasury Chambers,

Lord President,

1) The Royal Society have come forward a second time with a claim for closer co-operation between themselves... and the Government in the prosecution of the war...

I do not think they have any real ground of complaint but the setting up of a Committee...would I think satisfy the Society and at the same time strengthen the confidence of the public in that it would be a standing proof that Science is being given its due place in the scheme of National Defence, and would meet some of the ill-informed, if not ill-natured, criticism contained in the Penguin Special ('Science in War') which has had a large circulation. [The authors were anonymous.] I therefore strongly recommend your approval of the proposed Committee under Lord Hankey's chairmanship.

At the foot of this minute is written: 'I entirely approve these proposals. NC 23.ix.40'.

There follows a charming letter from Neville Chamberlain to A. V. Hill, dated 26 September, thanking him for his memorandum and saying:

No one is more conscious than I am of the importance of harnessing Science to our chariot and making her one of the team that will ultimately win this war. Indeed, you are pushing at an open door so far as I am concerned and ever since Sir William Bragg's letter reached the Prime Minister we have been examining his proposal and endeavouring to meet the Royal Society whose co-operation I know is most welcome to the Government.

For your private and personal information I may add that I have just made a recommendation to the Prime Minister on this subject which I hope he will see his way to accept.

But feelings were very different on the two sides of Mr Chamberlain's 'open door'; and if A. V. Hill had not threatened explosion, and if the Government had not been feeling so sore about the Penguin Special which had declared that war would never have happened if a scientist had been in charge of foreign policy, would WHB's carefully reasoned arguments have won the day?

This is Churchill's 'personal minute' agreeing to the committee:

> I see no objection to this, provided it is understood:
> (a) That the secrets upon which the various departments are now working, particularly those concerned with A.I., G.L., A.S.V., P.E., etc, shall not be imparted to a new large circle
> (b) That the time of the Scientists and Committees who are at present engaged in working for the Government shall not be unduly consumed.
>
> As I understand it, we are to have an additional support from the outside, rather than an incursion into our interior.
>
> [initialled in red ink] WC 27.9.40

It is hardly enthusiastic.

Lord Hankey was to be chairman of the committee. Barlow had written to invite him to take this on (18 September 1940) saying:

> I think it is clear that the Government will have to set up some committee of the kind, if only to keep the scientific people quiet, and it is conceivable that it might be of some use.
>
> It seems to me also that is must be nominally attached to the Lord President of the Council as he is the Minister responsible for the Government's relations to scientific activities in general. Mr Chamberlain, however, has no time to spare for giving any personal attention to the thing. . .if you could take effective charge nominally on his behalf, that would be a happy solution.

And the government probably thought it would also be a nice job for Hankey, who was feeling his decline from years of power as Secretary to the Cabinet.

Hankey's acceptance shows the same lack of eagerness for the Committee, but 'I will do my best to obtain some value from it [he wrote] without making unnecessary work for the Departments concerned'.

So then at last a letter was sent to rejoice the Royal Society's heart. Barlow wrote:

Dear Sir William, Sept 1940

I know you will be glad to hear that following upon our correspondence and a full consideration of the issues involved by Ministers concerned, the Prime Minister, on the recommendation of the Lord President of the Council, has agreed to the setting up of a Committee very much along the lines you originally proposed. The only difference is that instead of the Chairman being the Lord President, or some other member of the War Cabinet, he should be Lord Hankey who has always been so closely connected with the Government Scientific Research bodies and who has naturally more time at his disposal than War Cabinet Ministers whose time is already very fully occupied with meetings of the Cabinet itself and already existing standing Ministerial and Cabinet Committees.

The secretary of the Committee will be a member of the Cabinet Secretariat and the Committee will report in the first instance to the Lord President as the War Cabinet Minister primarily concerned with Scientific Research.

The terms of reference will be as follows:

(a) To advise the Lord President on any scientific problem referred to them.

(b) To advise Government Departments, when so requested, on the selection of individuals for particular lines of scientific enquiry or for membership of committees on which scientists are required, and

(c) To bring to the notice of the Lord President promising new scientific or technical developments which may be of importance to the war effort.

I hope that I may advise the Lord President and Lord Hankey that you yourself and your two colleagues of the Royal Society will be willing to act. I should add that in our consultations we have already taken the advice of the secretaries of the Government Research Departments who have welcomed the establishment of the Scientific Committee proposed.

I shall be grateful for an early reply as the Prime Minister is anxious to be in a position to make public the announcement of the setting up of this Committee.

Yours sincerely, Alan Barlow.

The Scientific Advisory Committee, often called 'The Hankey Committee', was set up on 8 October 1940, Chamberlain's last official action as Lord President. Sir Alfred Egerton, Secretary of the Royal Society and one of WHB's two colleagues on the new committee, sent my father the following letter:

9.10.40

My dear Sir William,

It is very satisfactory that the Committee has at last come into being... Even back early in 1938, when there were discussions going on about 'Social Aspects of Science' and various people put forward suggestions I felt there was lacking a focus point and suggested the RS might create one along with the other advisory Councils... It is in the reconstruction period that [the Committee] can perhaps prove most useful, and I hope it will remain as a permanent piece of organization... [He thought the Secretaries should start to form a register of laboratories; that research should go on wherever possible even if it was not directly concerned with war work because] one never knows what is going to turn up round the corner...

I want to say that if I ought to be replaced by someone weightier like Tizard would be, that I shall be perfectly happy because it has been my main aim that a Committee like this should be got into being and that it has come about gives me all the satisfaction that I want, for I am sure it is the right change in organisation.

For the RS is performing its function in leading on to new things, and not letting itself become a mere academy, which it might have owing to the great increase in scientific work carried out under the aegis of the Government for the State. We link the freedom of the Individual with the State.

Yrs ever Jack Egerton

Please forgive this scrawl. I wrote something yesterday, but, owing to air-raid inconvenience, have not been able to find what I wrote!

And, speaking of the 'Hankey Committee' in his last Presidential Address, December 1940, WHB proclaimed:

Thus a great opportunity is opened after long expectation; and the Royal Society is largely responsible for the development of that opportunity. . . We remember that it is science itself, not scientists, that we are trying to lift to the high places. . . We do not claim that scientists shall be entrusted with authority because they are scientists: we do claim that authority shall be exercised in the light of a knowledge which grows continuously, and with continual effect on politics, on industry, and on thought itself.

This would have made a fine ending to his Address, and to his Presidency; it is characteristic of WHB's thought that he did not leave the matter here, but returned instead to the education of the young as an essential part of the whole subject.

. . .Just as in former times schools and colleges were founded to train men for the service of Church and State, in ways which were appropriate to that high end, so now we have to see to it that the men are produced by our educational systems who can appreciate and act up to a new state of affairs. This can be done without jettisoning any of the fine instruction which has been a proud feature of our older systems.

I think that this is not essentially a matter of the re-arrangement of school time tables, or the building of scientific laboratories, though such tactical methods must have their due consideration. This is a personal matter, as has been the case with every great human movement. We have not to force the use of new tools, but to encourage and develop a new appreciation and a new attitude. Our best method, as ever before, lies in our own actions. If we, in the continually increasing contacts of scientists with public affairs, can show that we have something of great value to contribute, and that we give it freely, placing our individual interests below those of a greater purpose; if we try to understand the motives and principles of those whom we meet who may not see our vision just as we may fail to appreciate theirs, then by so doing we have the best chance of bringing about the changes that we desire. It is the personal contact of the scientist, especially with those who are charged with duties to the nation, that is the moving force. That is where these new associations of science with government may mean so much, and shall mean it, if our devotion can achieve its purpose.

It was with devotion on the part of the scientists and gathering interest on Hankey's part that the Scientific Advisory Committee got down to work. WHB served on it for sixteen months. Some account of what was accomplished and what happened to the Committee after WHB's death can be found at Note [27] to this chapter .

11

THE FAITH OF A SCIENTIST

Of science, WHB once said: 'Science is merely knowledge. It has no morality at all. If it falls into the hands of evil persons working for evil ends we, as scientists, can do nothing about it. To the question of whether Eve should have eaten the apple of knowledge, with all the ills that meal created, the pure scientist can only answer "Yes".'[1] This was said to a reporter after WHB had given the Pilgrim Trust lecture to the American Academy of Sciences in 1939.[2]

In the lecture he had said of scientists: 'Their achievements have value of one kind, and the spirit in which they worked has value of another kind; and the latter value is far more to be desired than the former. We may truly say of some of our greatest men of science that the world has gained more from their lives than from their discoveries...' Here he gave examples, Pasteur and the Curies, and Faraday, whose 'reverence for truth and unselfish devotion to its acquisition have a higher value than the laws which he established... In brief the spirit in which knowledge is sought and the manner in which it is used are more important, more real than knowledge itself.'

Talking to that reporter WHB placed science on the side of blameworthy Eve. The church has often blamed Eve; for centuries the church was blaming science for upsetting religious belief. But, five hundred odd years ago, a churchman wrote this:

> Learning itself is not to be blamed, nor is the simple knowledge of anything whatsoever to be despised, for true learning is good in itself and ordained by God... But because many are more eager to acquire much learning than to live well, they often go astray, and bear little or no fruit.[3]

Here there is no antagonism to learning in itself: Thomas à Kempis was just putting knowledge in its right place and at the same time giving a warning to the learned, including scientists; a warning often needed.

Thomas the monk was not afraid of knowledge; WHB, scientist, put the quality of life first. Their viewpoints join across centuries of conflict between science and religion.

His own resolution to the problem WHB gave, characteristically, in a lecture to children: all important things should be so well understood that they can be expressed simply, and he ended his lectures on 'The World of Sound' by saying:

> In our lives, in all that we work at and strive for, it is of first importance to know as much as we can about what we are doing. . .so that our work may be the best of which we are capable. That is what science stands for. It is only half the battle, I know. There is also the great driving force which we know under the name of religion. From religion comes a man's purpose; from science, his power to achieve it. Sometimes people ask if religion and science are not opposed to one another. They are: in the sense that the thumb and fingers of my hand are opposed to one another. It is an opposition by means of which anything can be grasped.[4]

WHB wrote this in 1920. By now, more than half a century later in a less confident world, the fierceness of controversy has faded and theology itself has changed. Most men care less, and some understand the scope of science better; for which understanding WHB battled all his working life, and his son after him.

WHB's approach to all living was the same: you had a great idea, you tried it out; if the results proved your idea then you used the conclusion as basis for the next advance. This was so in the laboratory; it could be so in any discipline from education to literature to cooking. It was so in religion. Religious faith to WHB was the willingness to stake his all on the hypothesis that Christ was right, and test it by a lifetime's experiment in charity.

WHB started on the experimental journey of his faith in the unquestioned religion of a country church where his grandfather Wood the naturalist preached, and his mother played the barrel organ. Later, at Market Harborough, church-going was equally a matter of course, but religion was less genial under his grandmother Bragg's narrow Anglican eye, and Uncle William's conception of duty and hard work. Bible reading was obligatory. WHB used to say: 'If I have any style in writing it is from having been brought up on the Authorised Version.' He knew his Bible; and could usually give 'chapter and verse'.

And so life went on through his school years with work, games, chapel and duty, until his last year at King William's College, when the wave of religious excitement hit the senior boys, devastating them and sweeping them off their feet. In WHB's own words in his autobiography:

> We were not singular, no doubt. Nerves are said to be liable to disturbance when boys are turning into men: and religious storms are common enough. Anyway, we had it badly, in the sense that we were terribly frightened and absorbed: we could think of little else. We had prayer meetings and discussions. We were told that if we sought we should find, and we hoped that somehow or other, at some time, we should suddenly be converted and know that we were saved, and avoid eternal damnation and hellfire. The issue was indeed quite simple. 'If we believed, we should be saved.' By 'saving' the words meant – so we thought – being delivered from an eternity in which we should be subject to pain worse than that of having teeth out without anæsthetics:[5] such an interpretation, if thoroughly absorbed, would, indeed must, make lunatics.[6] That the Christian world has gone on for centuries accepting that interpretation and yet going about its ordinary business is one of the strangest facts of history. A few have felt its force, and tried to save themselves by terrible acts of self-sacrifice of useless character, others by noble works in which self is forgotten: but the marvel is that there the words stand and are allowed to stand. The other word in the phrase is 'believed'. We took that to mean that we accepted the truth of all the statements made in the New Testament by Christ, by Paul and others, literally. (On second thoughts, I doubt if 'believing' meant to us anything so comprehensible: we were in fact not quite sure what it did mean: but it was something we had to do to be saved.) We were not sure we believed; sometimes we thought we did and thought we had attained to the 'peace that passeth all understanding', and any one of us who had got that far was the envy of the others. Then such a one would begin to have doubts about his believing, and would be in the soup again.

It really was a terrible year. If a boy came to me now, and told me he was in trouble as we were in trouble, I should tell him that believing and saving could not mean literally what we thought they meant: our intellectual difficulty was no more than an intellectual difficulty, due to the fact that words mean different things to different people. True, there are still many who take the words as literally as we did, but they do not really accept them, they put them in the cupboard and shut the door. In our time our masters and teachers did not dare to say that, or did not realize it. Therefore I have always felt strongly that it is necessary to be continually emphasizing the evil effects of 'literal' interpretation; when you come to think of it, 'literal' means only what the reader thinks that the writer meant, basing his interpretation on words which may have many values to as many people, and may change their average value during the passing of years.

Jack – my brother – and I once summoned up courage to go to the headmaster and ask for counsel. He was very friendly and sympathetic and we all knelt down in his study while he prayed for us. But he did not resolve our difficulty. The storm passed in time, by sheer exhaustion, and the fortunate distraction of other things, work and play. It lasted with me for some time after I left school, and then faded away. But for many years the Bible was a repelling book, which I shrank from reading.

The hysterical anxiety of that year would be particularly upsetting to WHB, who was so good, so kindly, but whose favourite atmosphere was a cheerful calm.

At Cambridge WHB got on with his work: WHB's advice to anyone in doubt or trouble was always to 'get on with the next thing'. Work was his own next thing.

Then Australia, success and new life. WHB was warmed by the friendliness of the Adelaide people; deeply impressed by the natural happiness of the Todd family and their creed of making other people happy and looking after them. Churchgoing was a serious and unquestioned part of Todd life. And so, within a few years, the young Professor married to Gwen Todd became churchwarden of St John's Church Adelaide; he also obtained a licence to preach. Childhood faith had only been cut down by the King William's College experience, now it was springing from the root again.

At that time, when babies were photographed naked on fur rugs and first curls were cut and preserved, Mother (GB) cut off a fine flaxen curl from her younger son's hair and must have picked up a sheet of paper from the waste-paper basket to wrap it in. The curl has survived; on the paper in which it is wrapped (headed 'University of Adelaide') are written WHB's rough notes, surely for a sermon. He directs himself to

> Read, paraphrasing: . . .
> Christ's rule and example showed God as our Father and us as His children, a society in which love governs all. Then if we seek a rule of conduct we should think of what we should like children to be like and what we should wish them to do. We like them to be hardworking, eager, cheerful, sympathetic. We like them to enjoy themselves thoroughly. We must be sad and in pain sometimes, but let us be happy as much and whenever we can, and whilst we are well and happy let us help all who are not. The more we strive to enjoy ourselves the more happiness we shall be able to communicate to others. For we trust that this life is a preparation: not a final probation. And we may be in training for something nobler. Some lives seem full of work, some empty: but so some fields are heavy with crops while some are fallow. Still, there is change going on in the fallow.[7]

For those who find the creed of happiness a bit ingenuous, I quote this sentence from one of Dostoevsky's letters: 'Men are made for happiness, and anyone who is completely happy has a right to say to himself "I am doing God's will on earth"'.

The value put on happiness reflects what Australia had given WHB; he had absorbed the Todd creed. There are no records other than the curl-wrapping to tell of WHB's thoughts on God and man during the Adelaide years. But WL could remember the Bragg family's practical contributions to their church; his father presented an electric blower for the organ and was much distressed when they forgot to oil it and it seized up; his mother designed a pattern of grapes and vine leaves and stencilled it round the interior of the church, and found church fêtes a delightful opportunity for getting up something in fancy dress. They remained staunch church members. When the family left for England in 1909, Professor and Mrs Bragg were given a farewell reception by the congregation of St John's; the Professor was presented with a set of large photographs of the church mounted on stiff card, and Mrs Bragg was given a green smelling salts bottle with a silver top.

In Leeds, the family attended service at Leeds Parish Church, and made friends with the Vicar, Samuel Bickersteth, and his family. Bickersteth once organised a religious public meeting in the Square in the centre of Leeds and asked WHB to speak as a scientist on the side of religion. Out of the kindness of his heart and friendship towards the Bickersteths WHB allowed himself to be captured and displayed. WL listened to his father's speech, miserable under the baleful comments of his neighbour in the crowd.

Soon I was old enough to accompany my parents to Matins. I licked the varnish on the gallery pew, and wondered if the Litany would ever end.

In London, after 1915, WHB would accompany GB to 'Early Service' as it was invariably called then, 'Communion' being a word too reverent to be easily used, and 'Eucharist' too High Church; less often he sat patiently and listened a little uncomfortably to a sermon. Never after his return from Adelaide did WHB take any active part in church affairs; increasingly he felt apart from organised religion. The Church's non-comprehension of the scientific point of view distressed him greatly.

But here and there some theologians had understood. How it would have cheered WHB to know that Stewart Headlam had declared in a sermon preached so long ago as 1879:

> Thank God that the scientific men have. . .shattered the idol of an infallible book. . .for so they have helped to reveal Jesus Christ in his majesty. . . He, we say, is the Word of God; he is inspiring you, encouraging you, strengthening you in your scientific studies; he is the *wisdom* in Lyell or in Darwin. . . It gives us far grander notions of God to think of him making the world by his Spirit through the ages, than to think of him making it in a few days.[8]

But in the 1920s antagonisms were still rife. The bitterness and suspicion of many in the Church were matched by the somewhat unimaginative scorn for religion of the lesser men of science, assertive in their rise to power. There was little understanding, great heart-rending; and much confusion brought to simple good people.

Scientists were accused of putting science in the place of religion. WHB denied it, and tried to explain: 'Research was not a religion, but the act of a religion. It implied a certain faith in the beauty and the purpose of the Universe. That beauty was meant to be seen by those who had eyes to see, and each further step extended the horizon.'[9]

Again and again he tried to define the scope of science, and the viewpoint of the research scientist. In 1928 he gave the Presidential Address to the British Association at Glasgow. At that time, one of WHB's great supporters at the RI was Sir Arthur Keith, Conservator of the Museum of the Royal College of Surgeons and a most courteous kindly man. But Keith was accused of holding a mechanistic view of the Universe, and his materialism was much quoted as a threat to the Church. WHB said in his address:

> Scientific research in the laboratory is based on simple relations between cause and effect in the natural world. These have at times been adopted, many of us would say wrongly, as the main principle of a mechanistic theory of the universe. That relation holds in our experimental work; and as long as it does so we avail ourselves of it, necessarily and with right. But just as in the case of research into the properties of radiation we use a corpuscular theory or a wave theory according to the needs of the moment, the two theories being actually incompatible to our minds in their present development, so the use of a mechanistic theory in the laboratory does not imply that it represents all that the human mind can use or grasp on other occasions, in present or in future times.

The address was reported in the periodical *America* in an editorial article headed 'Science and the Soul': in the same issue there is an article by a Jesuit who prefers WHB's viewpoint to Keith's but ends by saying:

> until [Bragg] speaks more clearly and more definingly on the

> questions of the soul and of religion it were well to leave off quoting him. . .our Faith and our philosophy need no such rickety shoring-up.[10]

WHB went on working for understanding; and in 1935 he accepted an invitation from the Bishop of Taunton to speak at the Diocesan Conference at Wells. His address was on the position and function of science; after his talk 'discussion followed in which members of the Conference and students of Wells Theological College took part'. WHB had taken great trouble with his address and had set off with hope; but he returned downhearted, rarely had I seen him so down-hearted. The theologians there, secure in their own point of view, had no wish to try and understand the scientist's viewpoint and he had been unable to get anything across to them.

The earlier accusation against science had been that science would de-throne God; the later accusation as technology advanced, was that science would also destroy man. In 1927 the Bishop of Ripon cried out fearfully 'Let us halt scientific research for ten years!' but WHB's replies were always variations on the same theme:

> Let us not fight uselessly against discovery and invention, but let us learn how to use the results.[11]

And again:

> Obviously, a new knowledge of tremendous power is steadily unfolding. No man of good intent can afford to neglect it. The better the men into whose hands the power falls the better the use that will be made of it.[12]

Again:

> I believe the greatest enjoyment of life is in behaving in a neighbourly fashion. You cannot do that effectively without knowledge.[13]

In 1938 he developed this further in his broadcast on Moral Rearmament quoted on page 110. Alas that will and heart seem to lag behind the means that knowledge gives; however 'Scientists', as WHB said on another occasion, 'are always optimists'.[14] Certainly WHB's hopes for how science could benefit mankind never flagged.

The Hibbert Journal was a journal of standing in the period. It had been founded 'to promote the spread of Christianity in its most simple and intelligible form, and the unfettered exercise of the right of private judgement in matters of religion'. In 1938 the Editor persuaded WHB to write an article for the Journal, to try once again to keep the mechanistic theory in its place. The article was to be a response to a plea by Sir Richard Tute that scientists would 'denounce the false materialism which masquerades in the name of scientific truth'. Tute had written that the false materialism was 'a view of life that regards man as a product. . .of a mechanistic universe and the Universe as a product of blind forces operating through æons of time on dead matter'.

WHB proclaims:[15] 'I agree entirely with Sir Richard [Tute]'s condemnation of such materialism' and goes on to say that there was no evidence to prove that life and the Universe could be made by force on dead matter. He illustrates the limits of the mechanistic theory with a nice analogy:

> Suppose that a motor car were a novel object, and its driver hidden. An enquirer might begin an investigation of its nature and action. There are wheels which go round. Did the separate molecules of rubber and metal come together by chance to make the wheel? Men ask such questions in respect to the universe. Or was the assemblage due to the intelligent action of each molecule, or of each atom in each molecule? Or did intelligence arrive when the assemblage was complete, so that each wheel was alive though the molecules were not? Examination in the laboratory shows that the rubber behaves like ordinary rubber and the metal like ordinary metal. . .[and WHB imagines the inquirer testing other parts]. In the end he will come across the driver, and will now say that he has an explanation of why the car moves and acts with seeming intelligence. . . But in the case of the living organism, no one has yet come across the driver by any enquiry into the mechanism of the organism. It is also the case that at no stage of the enquiry has any action or process been found which transgresses the rules found by laboratory experiment. . . It is in this sense that the mechanistic theory of the universe is right, so far as we know at present.

But under the mechanistic rule:

> what becomes of free will and the moral struggle? . . . We may feel that the freedom is incomplete. . .still, we possess

> free will. It is an experience.[16] But mechanical sequence and free will are apparently in irreconcilable opposition. To the student of experimental science the position is not unfamiliar.

Again he cites the example from the seeming opposition of the wave and corpuscular theories; this instance always sprang to mind, written on his heart since his own controversy with Barkla over the two theories of the nature of X-rays.

WHB died in 1942. The year before, he had been up to Newcastle to deliver the Riddell lecture before the University of Durham. The title of the address is 'Science and Faith'. It was his last public statement of the faith in which he had worked through a long life. It finishes thus:

> It is time now to come to an end of what I am saying. I have not attempted to put forward any solution of the great problems of existence and purpose. I am indeed anxious to assert that in my opinion science does not suggest solutions. The scientist gathers knowledge of the world and of ourselves and as scientist stops there. As a man among other men he draws on the accumulating store. When he records his observations he adds to the book which anyone may read. Whoever reads it, himself or any other, is responsible for the use he makes of it. The scientist is responsible only for its accuracy, so far as that is attainable, and for its enlargement. Amendments and enlargements do not destroy what has already been written; on the contrary, after removing mistakes they enrich it. The scientist is always aware of a growing world of knowledge, ever presenting new and surprising things, rich in suggestion of what is beyond the vision of his time; he is continuously urged on by the fascination of the search. I believe also that he is conscious of a growing power of understanding, so that a generation may walk firmly where its predecessor moved haltingly. He becomes chary of prophecy and assertion, having found that his hypothesis can only be provisional, and that he never knows in full. Not only his facts but his appreciation is incomplete. But it is always in his power to experiment for himself, and there lies his happiness and his strength, his training and his conviction.

It is right to consider whether this attitude, acquired by experience, has any reaction on the religious life. I am not sufficiently informed to know how all types of mind are affected by the demand for the absolute acceptance of definite items of faith as a preliminary condition to progress. I believe I can safely say that to many minds this is an impossible demand. Conviction of the truth of any faith, so far as a man can measure the truth, is to be gained by practice, and it is here that the scientist finds an illustration in his own work. Every man, in the circle in which he finds himself, it may be a small circle, his means may be small also, can try the Christian way, and discover for himself and acquire his own convictions. He tests his faith. He has ever in front of him the hope that he will by doing his service play his part in binding the community together. That is his hope. As to the actual mode of the experiment, I will say nothing. We all know it well already: it has been enshrined in a thousand testimonies; it has been displayed in countless lives; it is all included in the lovely words of St Paul, simple though they are: 'And the greatest of these is charity'.[17]

12

THE MAN WHO WAS A SCIENTIST

In the beginning of this book I posed some questions about scientists. Through ten chapters I have enjoyed describing my father, hoping that some answers would emerge from the pages, even though I have been writing about the life of only one scientist among many. Now I near the end of the book I wonder if those grand sounding questions have been answered, even in part.

Let me start with the third; surely that question has answered itself in my father's case. Could any man have shown more responsibility to the Body Politic?

The second question is in two parts; to start with, how does the scientist come to make his discoveries? Here my father was not typical. I suspect most scientists nowadays worry at research from the beginning of their careers because research is expected of them. WHB only made his start on research in middle life, and then only because he had to prepare a paper. Not that he had not been keenly interested; he had just been too modest to enter into the field: which brings us to the second part of the second question. What is the scientist's urge? WHB's urge, so late to break through, was an urge he had in common with many true scientists, I believe; a divine curiosity; a wonder and delight in uncovering some new corner of the Plan. Here let my brother speak: 'the spur to the devotee of science is his belief that he has as his judge not "broad rumour", to quote Lycidas, but "the pure eyes and perfect witness of all judging Jove". It is as if, in the restricted field of natural philosophy, we are allowed to attempt a measure of absolute certainty, a glimpse of truth herself.'[1]

And now to deal with that first question, 'what sort of man is a scientist?' This whole book has been about the man WHB, yet I am sure I might have described him better: I have tried to make a portrait of him, but it is hard to convey his likeness. I have said that he was good and was good-looking; that he was modest; that he was kind. It does not sound arresting. It is easier to catch the likeness of someone with a large nose or a striking eccentricity: and I am hampered because I am a daughter, and must be thought partial.

How can I describe him further? It is said a man is known by his friends. I have thought along this line, but it stops almost as soon as it starts, for I do not think that WHB sought close friendships: he regarded the world with benevolence, enjoyed friendliness, and walked rather independently, if you rule out close family. He seemed not to require intimacy; he was, I think, too shy to make intimate friends.

There was one great friendship he did have, with Rutherford; but for a long time it was only a scientific friendship. Though they had met when Rutherford had called in at the Adelaide laboratory on his way to England in 1895, the friendship started when WHB wrote to Rutherford about his discoveries in 1904. During the following years they kept in close contact, writing warmly, page after page; one letter from WHB is thirty-four pages long, and their letters were frequently of six or eight pages written in terms of great friendship – but only about their work. There is no reference to family life until 1907 (three years after the correspondence began) when WHB was considering going to Montreal, at Rutherford's instigation. WHB wrote, 29 August 1907, that 'The most serious difficulty I have to face is that of the moving of my household. I must tell you that I married here about 18 years ago, and that we have two boys, one at the University and one at school; and a little girl six months old. We have built our own house: and altogether have struck our roots very deep.'

Kinship of spirit is another thing, and can be felt across the generations. WHB often expressed his reverence for Michael Faraday; I believe they would have felt akin. I do not compare them in magnitude of scientific achievement, but in outlook and character. Each had the same humility of thought, the same awe before Creation; both tried to impart their vision and enjoyed especially, I think, sharing their enthusiasm as they lectured, and especially as they lectured to children. Both helped their country; each remained unsophisticated, simple to his end.

WHB wrote this of Faraday's great Diary: 'The diary is far more than a mere record of work done. It is a most human story of the reasonings and researches of one of the greatest experimental philosophers written by the man himself as he would have written to a friend.'[2] It was certainly as a friend that WHB used to read it.

Let the comparison rest there, for WHB had children, Faraday had not, and the gentle retiring Mrs Faraday was very different from GB.

Perhaps WHB's attitude to his children can explain something more about him. It was based on this, that they must be free; absolutely free; he could not bear to sway them by his wishes. When important advice was being sought he grew very unhappy. He would shift in his chair, grunt sympathetically saying nothing, get up, try and change the conversation, until at last, worn down (often by WL on a flying visit to the parental home) he would say 'let me think about it'; and then, after a day or two, would send a letter of carefully reasoned advice, reasoned with exquisite regard for the opposite point of view.

Once, soon after the First World War, WHB wrote to WL to ask him if he would care for a job with the General Electric Company at £1,000 a year. WL wrote back with some astonishment, 'I had never contemplated anything else than a University career' – nor, I feel sure, did his father want anything else for him. Yet WHB, asked to find a man, had set out the advantages of the GEC job, to be fair to his own belief that it was to the country's advantage that good scientists should go into industry, a rather new idea at that time. A very different example of the same careful neutrality was provided when, as a girl, I consulted WHB on a very important matter. Leaning over the gate together I asked him 'Shall I. . . ?' and he replied, 'My dear, you can marry a blacksmith if you like.' I don't think he would have been best pleased if WL had become an industrial physicist, or if I had married a blacksmith; but in each case he was just trying to leave his children entirely unhampered by his own point of view.

We both turned to him for advice; and Alice, his daughter-in-law, would write for advice when anxious about an over-worried husband. Sometimes one would have liked to be more firmly guided. GB of course always knew exactly what one ought to do, and said so; perhaps that was worse for us.

Undemanding, self-effacing, are two more adjectives that apply to WHB. My husband's memories are worth quoting here since he shared two homes with my father (the flat at the RI and our country home, Watlands) from early in 1934 until WHB's death in 1942. To live in an older man's house and take a share in its domestic running and finances is not necessarily easy for a newly married man, but my husband cannot remember a single occasion when this position caused him any frustration or awkwardness; indeed his greatest difficulty was to remember not to express a wish for any project before discreetly probing to find WHB's views; for any expressed wish was sure to be granted if humanly possible. The two men shared their home life with a comfortable trust and affectionate reserve.

How was it that so modest a man, a farmer's son who till the age of forty-two spent his life as a conscientious teacher in a small university on the other side of the world (a world not yet linked by radio or plane) – how was it that, returned to England, he became in a few years' time a spokesman for science: later, *the* spokesman one might claim? Perhaps the answer lies in that long and happy exile; perhaps those twenty busy, happy years in Australia were as valuable to him as desert years to a prophet, a time of unhurried preparation. He had a long way to travel from his simple beginnings; Australia gave him time to grow slowly and strongly, and to gather confidence intellectually. He had time to think out the principles that guided his life, to order his thoughts; long practice in teaching and exposition. Consequently, when later in life he was back in England among problems at the centre and in the urgency of war years, he forged towards his determined goal, not without any doubt of the way, but with clear principle; so that his life was rather like his manuscripts, well thought out, written almost without crossings out.

Some idea of the character who died in 1942 can be gathered from obituaries and letters. Letters of condolence only dwell on the best, of course, but something can be distilled from them: one notices how the same adjectives recur; one is struck by the warmth of affection for WHB.

There were so many letters and telegrams; a telegram from the King and Queen, a letter from the head warden of Firewatchers Block 6A; letters from the Archbishop of Canterbury, Hankey, R. A. Butler, from science school teachers, and the man who looked after our car at the garage; from the scientific community all the world over. An RI porter said he felt he had lost a father, and research people from the Davy Faraday that they had lost the head of the family. Again and again the words kindness, modesty, gentleness recur, and the phrase 'he had the humility of a great man'.

WL received a letter from Charles Glover Barkla, the old opponent with whom WHB had argued so fiercely, saying: 'Although I never saw much of [WHB], it was impossible to meet him without feeling something of his kindness of spirit.'

From Sir Henry Dale: 'I have known a number of the great ones but none of his stature, I believe, who was beloved with quite so general a warmth of affection.'

And Clement Attlee wrote: 'For the last two years I had been seeing him very frequently about the Scientific [Advisory Committee on Food] of which he was chairman, as I was chairman of the Food Committee of the Cabinet. I admired very much his devotion to the public good and his invariable kindness and courtesy.'

It was in this public character of science that WHB the scientist made such contribution as a man. 'How can science help?' was a motive throughout his life, from the days when he tried to get Australian school children better taught to the days of the Second World War; and then, it was not so much what he did that brought scientists into Government positions, but rather what he was that led Government men, who knew so little about science, to accept his word that scientists and science were 'good things' for the country.

A. C. Egerton, Secretary of the Royal Society, and WHB's ally in getting the Scientific Advisory Committee started, wrote to WL: '[WHB combined] real humility. . .with richness of mind and spirit. . .[we sometimes] had rebuffs from the official world, but it was your father's steadfast refusal to be prevented from achieving what was right that brought about the clear relation between Science and Government which now in some measure exists.'

Strength and gentleness; add to this lucidity in exposition; he lectured to children with charm, he could explain to the official world. And he had no personal ambition: it was Science, not scientists who must be raised to power. Personal power politics among scientists distressed him greatly, they smudged the face of science. He always tried to take the kindliest view of every man he met; I only remember one scientist who made him wince, and a few persons with whom he was uncomfortable. Occasionally, sitting quietly of an evening, he would draw in his breath and give a small shiver: some memory had come to the surface like a bubble on a quiet pool; he had suddenly remembered somebody or something which hurt, or somebody he thought he had hurt, perhaps many years ago. And I, sitting opposite, would tell him he should tear up those old bad memories, as one used to tear up a receipt after seven years.

I have already quoted from Professor Andrade's account of WHB's scientific work, and from W. T. Astbury; I only want to quote here one more scientific verdict, from J. D. Bernal. Bernal worked for some years in the Davy Faraday Laboratory, and used to agitate us rather because he was so very pale in face and red in outlook. He wrote to WL saying:

> He [WHB] was in a way, scientifically, my father too. He took me up and helped me through all the critical stages of my career. I was always proud of having worked under him. I had always meant to tell him how grateful I was, but somehow one never does and then it is too late.
>
> I was specially fortunate in the time that I came to the RI, the beginning of the second great period of his work. Less exciting perhaps to him and to you than the discovery ten years before, for me it was far more than I could have hoped for. In those five years the outlines of all the crystal types were laid clear by his work and yours. Since then we have only been filling in details.
>
> I remember how exciting it was: – the change from the large to the small metal ion, the ideas of packing, the break into organic chemistry – and how he planned the attack and was as excited himself as the youngest of us.
>
> He was a great research chief, encouraging and interested without interfering and knowing how to get the best out of us and when to send us out to fend for ourselves. I learned all I know about how research should be organized from his example. . .
>
> There must be few men in the world whose work was so successful, so important and so unique. His name will live in the structure of matter as sure as Galileo in the heavens, or Faraday in electricity.

'His name will live as Faraday's' – I think that was affectionate enthusiasm on Bernal's part. But WHB's name will live, joined with his son's; they have their place happily together in the history of Science, though it was not always completely happy in their lifetimes – people were always mixing them up, causing public confusion and some private heart-burning to WL. A letter from WHB to his sister-in-law Lorna Todd in Adelaide illustrates this. WHB wrote delightedly on 5 January 1941:

> you will learn by newspaper cable that Willie is knighted. Isn't that fine? . . . He will have to be Sir Lawrence: we can't have confusion worse than ever. I am so very glad for his sake. In spite of all care, people mix us up and are apt to give me a first credit on occasions when he should have it: I think he does not worry about that at all now, and will never anyhow have cause to do so now. I think I am more relieved about that than he is.

But though WHB wrote so hopefully in 1941 I do not ' ink WL completely lost his reserve about his father until, in the years before his own death, his gathered wisdom and success and humour had dispelled the final shred of cloud that had hung between them; and at last he felt able to plan this book, but did not live to write it.

By then the pendulum of public opinion had already swung, and a symposium of X-ray crystallographers gathered from all over the world at the time of WL's eightieth birthday, hailed WL rightly as the father of their new science: but they hardly mentioned WHB in their speeches. I, listening, felt puzzled and a little sad. I took an opportunity to ask Sir George Thomson, my brother's best friend and fellow physicist, how he regarded the situation. He wrote in answer (30 December 1973) saying 'I do not think it helps to guess who deserves the most credit. In such circumstances – even if the two people meet one another only *once* – it is impossible to tell who really pushed the cart over the hill, I think it is a mistake to try.'

Now that both lives are over one can see how easily all the confusion came about: they were so different in temperament, yet their work and aims were so alike. Consider: they shared discovery, their research was parallel, and though Cambridge and Adelaide cannot be compared, both were Professors in the English industrial north, both Directors of the RI organising research teams in the Davy Faraday Laboratory. In war both worked as scientists for the armed services. Both had artistic skill in exposition (and some indeed in painting), and WL carried on his father's crusade for education with his lectures to school children at the RI. The aim of both was to explain the province of science, the father during the optimistic years of science's rise to power, the son in the years of disappointment that science had not answered all the world's problems. And if the outcry (with which WHB fought) that science was destroying religion had largely died down by WL's time, how hard WL tried to get the province of science understood, writing and talking on 'The Spirit of Science'; 'I've explained it so often' he would say.

To each, science was an art, research an adventure, and life an experimental journey which they lived with enthusiasm. For each, life was physically hard as he neared eighty, but neither grew old. And each might even have been 'Sir William'.

Future generations remembering them will call them 'the Braggs' and speak of 'the Bragg contribution' to science in the twentieth century. WHB's life is only the earlier part of the story.

During his father's lifetime WL was the 'son of his father'; now in the 'seventies WHB has become a rather dim figure, the father of his son. Yet WHB's scientific writing is still cited as often as Rutherford's, and much of what he had to say as a man applies now as forcibly as when he said it, and most appositely for this generation obsessed by the problem of communication.

In one of his last published articles he looks at the differences in modes of thinking amongst various types of people. 'Can we bring these very different thoughts into harmony by amending and enlarging the knowledge on which they are based? Can we bring it about that the politician and the man in the street, the theologian and the man of letters, the industrialist and the soldier, and the scientist himself, are so well informed that their thoughts are interchangeable? Can we induce each type of man to speak with sympathy for the various conditions and minds of those whom he addresses?'[3]

To accomplish this was WHB's vision and the goal towards which he marched steadily and hopefully. WHB's way was through science; but to seek the unity behind diversity in nature and to link men's minds were to him two parts of the same quest.

NOTES

A bibliography of William Henry Bragg's publications was compiled by Dame Kathleen Lonsdale and printed in *Obituary Notices of Fellows of the Royal Society*, 4, 292 (Nov. 1943). See also *William Henry Bragg and William Lawrence Bragg: a bibliography of their non-technical writings* (Berkeley: University of California, Office for History of Science and Technology, 1978) with an introduction by J. L. Heilbron.

CHAPTER 1: pages 1 to 3

[1] Sir Lawrence Bragg FRS and Mrs G. M. Caroe (Gwendolen Bragg), 1962. Sir William Bragg FRS (1862–1942). *Notes and Records of the Royal Society of London*, **17**, 169.

[2] Sir Lawrence Bragg, 1975. *The Development of X-ray Analysis*. London: G. Bell & Sons.

[3] W. H. Bragg, 1932. Commemorative Oration on 21 September 1931 at the Queen's Hall. *Report on the Faraday Celebrations, Royal Institution of Great Britain.*

CHAPTER 2: pages 4 to 28

[1] The manuscript autobiography of William Henry Bragg is short; it was written in 1927, with notes added in 1937. It was not written for publication and has been included here in an edited version. Professor Andrade used parts of it for his Royal Society obituary notice: *Obituary Notices of Fellows of the Royal Society*, **4** (Nov. 1943). A copy of the typescript is among the Bragg papers in The Royal Institution.

[2] Lloyd's Register for 1852 records that 'Nereids 466 tons', owned by Bushby, was LOST on a Destined Voyage Liverpool–Calcutta. The Master's name, given as Mecquels, was really Michael. 'John Michael' Master of Nereides, 530 tons, signed RJB's apprenticeship indenture in 1846. The ship had been built that year in Workington: Captain Michael was taking on a new ship as well as a new apprentice. Shipowners called Bushby appear in the Liverpool records until the end of the nineteenth century.

[3] William Henry Bragg went up to Trinity College Cambridge in 1881; his elder son William Lawrence Bragg in 1909; WL's elder son Stephen Lawrence Bragg in 1942. All held mathematical scholarships, and the Master, Sir J. J. Thomson, wrote to congratulate WHB, pointing this out as a family record, greatly pleasing him. A fourth generation has continued the sequence.

[4] WL noted WHB's debt to Uncle William as follows:

> In 1897, when I was seven years old and Bob [five], the parents made the momentous decision that my father should take a year off and the whole family should spend it in England. I think the main

> object was that Uncle William should see my father again and his family. . . My father's successes in school and at Cambridge had been a source of pride and joy to Uncle William, and it was a great blow to him when my father left England for the post at Adelaide University. My father felt he owed everything to Uncle William and must see him again as he was getting to be an old man. *W. L. Bragg, MS. Autobiography; Royal Institution Archives.*

[5] Charles Henry Rendall (born 1856) was a close relative of Alban Caroe, whom the author married many years later.

CHAPTER 3: pages 29 to 51

[1] The book referred to is Augustin Privat-Deschanel's *Elementary Treatise on Natural Philosophy* translated and edited, with extensive additions, by J. D. Everett. It was in four parts, first published in London between 1870 and 1872. The 1882 (sixth) edition was probably the one WHB read.

[2] 'In the days of my youth', *T.P.'s and Cassell's Weekly*, 3 April 1926.

[3] 'A page from my life', *The Graphic*, 12 March 1927.

[4] *South Australian Register*, 11 July 1895.

[5] Cuttings from South Australian newspapers, 18 June 1896.

[6] Stanley Addison, who was WHB's technical laboratory assistant from 1899, prepared a colourful account of this time upon which was based a film made for the Council for Scientific and Industrial Research (Australia). Here is an excerpt.

> There is a report extant that one day in 1896 William Lawrence Bragg, the five year old son of Professor Bragg, of Adelaide University, fell from his tricycle and injured his elbow. When a Doctor who had been called could not decide whether a bone had been broken, the Professor told him to carry the boy to the basement, where the question would be answered.
>
> In the area below the laboratory proper the Professor had set up a primitive X-ray apparatus based on the discovery which Dr Roentgen of Wurzburg, Germany, had announced to the world a few weeks earlier. Although Professor Bragg's machine [had been] hurriedly brought into being, [it] nevertheless worked, despite its crudity, and was put into motion. Fascinated the Doctor watched as the big induction coil buzzed loudly, electric sparks crackled and vacuum tube emitted a weird green glow. Later on there appeared on the screen an X-ray photograph which distinctly showed the extent and location of the injury to the boy's arm.
>
> This was the first recorded surgical use of the Roentgen Ray, as it was then called, in Australia. Roentgen's discovery of X-rays was received with ill-informed criticism all over the world. Many members of the clergy protested against what they called 'the revolting indecency' of the invention. In the belief that Peeping Toms armed with Dr Roentgen's machine would prowl the suburbs, peering through brick walls and the like, some extremists demanded the Ray be outlawed and that the inventor's written works on the

subject be burnt. The records show that one London firm of outfitters advertised 'Ray proof underwear', while the State of New Jersey USA introduced urgent legislation to prevent opera glasses being fitted with X-rays and used for long distance invasions of privacy.

In Adelaide however, Professor Bragg was so interested that he built a machine of his own, and that accomplishment was the turning point in his life.

[7] From type-written account by Lorna Todd in the author's possession.

[8] 'Ether' was supposed to be the medium through which light and wireless waves were propagated. The fact that light passes through a vacuum seemed to entail the pre-supposition of ether, if one accepted the wave theory of light – the waves had to be waves *in* something; you could not have waves in nothing. Ether also helped to explain action at a distance, through gravity, magnetism, electrostatic attraction, chemical affinity. Scepticism about the existence of ether grew after the famous Michelson–Morley experiments of 1887. Fitzgerald, Lorentz and Einstein gave mathematical explanations of these experiments which made further discussion of physical ether unfruitful.

[9] *The Critic*, 16 October 1907.

[10] Cutting from a South Australian newspaper 18 June 1896.

[11] *South Australian Advertiser*, 16 October 1905.

[12] WL in his unpublished autobiography records coming on a trip to England in 1897 (see Note 4, Chapter 2). He describes Uncle William as follows:

Uncle William was the complete bachelor [he was in fact a widower], into whose life women and children had hardly come at all. I remember that Bob [WL's brother]and I demolished most of a rockery to make a fort in one corner of the garden under a large elder tree, and any annoyance felt by Uncle William was mastered by his amazement at our imagination, and at the ant-like perseverence with which we had moved, inch by inch, stones almost our own weight. He did, however, take the precaution of keeping his beehives in the strawberry bed. He was rather a vulgar old man – I remember his pinching us behind and saying our bums were like little cheeses. But even we realised his powerful personality, and stood in great awe of him.

[13] *The Daily Telegraph*, 10 January 1938. It is interesting to compare this remark of WHB's with the comments of Arthur Smithells in his Presidential address to Section B, British Association, 1907:

The perplexities of chemists at the present day come. . .in particular from the sudden appearance of the subject of radioactivity with its new methods, new instruments and especially with its accompaniment of speculative philosophy. There is an uneasy feeling that developments of great importance to the chemist are being made by experiments on quantities of matter of almost inconceivable minuteness. . .This reduction in the amount of experimental materials, associated as it is with an exuberance of mathematical speculation of the most bewildering kind. . .is calculated to perturb a stolid and

> earthy philosopher whose business has been hitherto confined to comparatively gross quantities of materials and to a restricted number of crude mechanical ideas.

By 1904 there had been nearly nine years of discoveries that (unlike the radio waves detected by Hertz in 1887) could not be reconciled with the accepted ideas of 19th century physics and chemistry. They were started by Röntgen (X-rays, 1895) and Becquerel (radio-activity, 1896) and were continued by the Curies, Rutherford and others, linking as they developed with the work on electrical discharges in gases started earlier by J. J. Thomson.

[14] 'In the days of my youth', *T.P.'s and Cassell's Weekly*, 3 April 1926.

[15] *The Register*, 13 April 1908.

[16] *The Register*, 12 January 1909.

[17] From leading article in *Brisbane Daily Mail*, 13 January 1909.

CHAPTER 4: pages 52 to 68

[1] *The Yorkshire Post*, 18 February 1932.

[2] Rutherford's letters quoted here and in the previous chapter are in the Royal Institution archives together with photostat copies of WHB's letters to him. The originals of WHB's letters are in the 'Rutherford correspondence' in Cambridge University Library.

[3] Evening discourse by WHB to the British Association 1912.

[4] Sir Lawrence Bragg, 1975. *The Development of X-ray Analysis.* London: G. Bell & Sons.

[5] W. H. Bragg, 1912. *Studies in Radioactivity.* London: Macmillan.

[6] W. D. Caröe, the architect, was not known to the Braggs in 1911 but later became the author's father-in-law. Having won the competition for the University College of S. Wales and Monmouthshire at Cardiff, he was appointed architectural Assessor for the new university in British Columbia. He knew Sir Richard Glazebrook because he had designed one of the first new laboratories at Teddington for Glazebrook as Director of the National Physical Laboratory.

CHAPTER 5: pages 69 to 78

[1] Sir Lawrence Bragg (1961) 'The Development of X-Ray Analysis', *Proceedings of The Royal Institution*, 38, 530.

[2] The dates of publication of these six crucial papers were, in chronological order:

W. H. Bragg, 1912. X-rays and Crystals, *Nature*, Lond. **90**, 219 (24 October 1912).

W. L. Bragg, 1912. The Diffraction of Short Electromagnetic Waves by a Crystal, *Proceedings of the Cambridge Philosophical Society*, **17**, 43–57 (11 November 1912).

W. H. Bragg, 1912. X-rays and Crystals, *Nature*, Lond. **90**, 360–1 (28 November 1912).

W. L. Bragg, 1912. The Specular Reflection of X-rays, *Nature*, Lond. **90**, 410 (12 December 1912).
W. L. Bragg, 1913. X-rays and Crystals, *Science Progress*, **7**, 372–89 (January 1913 issue).
W. H. Bragg & W. L. Bragg, 1913. The Reflections of X-rays by Crystals, **1**, *Proceedings of the Royal Society* A, **88**, 428–38 (17 April 1913).

CHAPTER 6: pages 79 to 92

1 WL's research on sound-ranging evolved a practicable method of locating gun emplacements by calculating the distances travelled by a shell's shock wave before being received along a series of field microphones. A similar method was later developed for submarine location by WHB; see p. 90 and *Journal of the Royal Naval Scientific Service* (July 1965), **20**, No. 4, 35. In his presidential address to the British Association at Glasgow, WHB described his son's work on sound-ranging, though he mentioned no names.
2 S. W. Roskill, *Hankey: Man of Secrets*, **I**, 140–1. London: Collins.
3 Roy M. Macleod & E. K. Andrews, 1971, 'Scientific Advice in the War at Sea 1915–1917: the Board of Invention and Research', *Journal of Contemporary History*, **6**, 3–40.
4 Op. cit. note 1 above, p. 19. This is the A. B. Wood Memorial Number of the Journal. Some correspondence between A. B. Wood and WHB has survived and was given by Mrs Wood to the Royal Institution.
5 *The Saturday Evening Post Magazine*, New York, 4 March 1916.
6 In all, WHB's group spent more funds, and succeeded in turning out more useful research, than the rest of the BIR put together. Expenditure up to 31.8.1917 amounted to £3,618 for WHB's submarine work; the whole of the BIR expenditure up to that date amounted only to £6,508 (MacLeod & Andrews, 1971. *Journal of Contemporary History*, **6**, 24). Yet how small was this sum compared with the amounts spent in the Second War.

CHAPTER 7: pages 93 to 109

1 Sir Arthur Keith, (1950). *An Autobiography*, 461. London: Watts & Co.
2 From the early years of this century until WHB came, numbers of members and quality of membership fell dramatically. When the London Institution (comparable to the RI, founded by Sir Frederick Baring and other City merchants in 1805) went into serious decline around 1912, there was some fear that the RI would go the same way. Crichton Browne held the fort, however, and the Duke of Northumberland's active Presidency helped to cover over the fact that Dewar's own administration of research was very weak. Dewar himself had been a first-rate scientist and administrator in his younger days; only age had changed him. In 1914 the situation became worse when the RI, at Dewar's instance, refused to allow the War Office to take over the laboratories. Correspondence in *The Times* and *The Standard* indicates some hostile reaction to what was (perhaps unfairly)

considered to be an anti-patriotic attitude. The RI made only small contributions to the First World War, and afterwards was nearly deserted. The Davy-Faraday Laboratory (founded by Ludwig Mond for the RI in 1896) was kept almost vacant. The Managers were said to be 'saving it for Dewar's successor'.

WHB's coming was the beginning of a long-sustained revival – a revival temporarily checked after WHB's death in 1942 by difficulties during and after the war, but renewed by his son WL eleven years later. The RI is now one of the few remaining wholly private scientific research institutions; but as its present Director, Sir George Porter, has noted 'it has produced more fundamental scientific break-throughs per square foot' than any other establishment in the world. Eight Nobel laureates have been Professors there.

[3] The *Lancet*, 14 April 1934.

[4] See W. J. Green's account of the explosion, *Proceedings of the Royal Institution*, 1956.

[5] The London County Council ordered that the Theatre of the Institution should either be closed or reconstructed with separate emergency exits. The total cost of rebuilding was put at nearly £80,000.

[6] The manuscripts were sold for £30,000 by WHB. They were not really so important as the newspaper and parliamentary critics made out, though in any normal situation WHB would never have thought of parting with them. They had already been carefully catalogued by the Historical Manuscripts Commission in the 1890s, and the most important material in them had been published. The dealer who bought them sold them later to Rockefeller for $500,000 (then about £100,000), but Rockefeller gave them to the US Nation. When Queen Elizabeth II visited the USA in the 1950s, President Eisenhower presented the papers to her as a gift. They now reside in the Public Record Office, where they rightly belong.

[7] The Royal Society was at Burlington House until 1967, when it moved to Carlton House Terrace.

CHAPTER 8: pages 110 to 122

[1] *The Times*, 19 October 1932.

[2] Broadcast Talk: October 1938. *Crisis Booklet No. 3*, S.C.M. Press 1938.

[3] During the early years of the Second War, the RI allowed the use of its laboratory facilities for some research into aeroplane wing stress, and for analysis of foods, fuels, ceramics and fabrics.

[4] Block 6A. The head was Mr P. F. Hubbard of Aspreys, the jewellers.

[5] Both on 'The diffuse spots in X-ray crystal photographs', *Proceedings of The Royal Society* A **179** (1941), 51–60, 94–101.

[6] Published under the same title by The British Council in 1942.

CHAPTER 9: pages 123 to 132

[1] E. N. da C. Andrade, 'William Henry Bragg 1862–1942', *Obituary Notices of Fellows of the Royal Society*, 4 (Nov. 1943) 277.

[2] Obituary Notice by W. T. Astbury, *Nature*, Lond. **149**, 28 March 1942.

[3] See note on bibliography of W. H. Bragg's publications, p. 179.

[4] See Chapter 5. They had already written a joint book in 1915, *X-rays and Crystal Structure*, London: G. Bell & Sons. From 1933 onwards they co-edited a series of volumes with the overall title *The Crystalline State*, London: G. Bell & Sons.

[5] It was not WHB's first experience of sorting out lines of research. It is interesting to compare this time with that in 1905 when he was sorting out work with Rutherford. He had been corresponding with Rutherford in Montreal at great length after his break-through into research in 1904. He wrote to Rutherford, 16 July 1905:

> It is a very generous thing to offer to keep away from the line of research which I propose to take... Of course I would be glad to go on with the work I am doing now, and to do so without fear of clashing with you. But I am sure you will understand me when I say that I could not let myself be in the way and I am very happy that I have been able to help a little bit.. .I should like best to work on some arrangement which has your approval: and therefore tell you what I propose: my idea at present is to push on with the investigation of the law which I have discovered, that the stopping power of an atom is nearly proportional to the square of its weight...
> My general idea is to attack the question of molecular and atomic structure by examining the absorption effects of the various atoms.

The feeling is of the junior looking up to the senior in science, although WHB was the senior in years.

[6] Sir William H. Bragg, 'Chemistry and the Body Politic': 7th S. M. Gluckstein Memorial Lecture 1935, *The Institute of Chemistry of Gt. Britain and Ireland.*

[7] Sir William Bragg, 'Research Work and its Applications': address delivered at the Sir John Cass Technical Institute on 30 January 1924, *Nature*, Lond. 1 March 1924.

CHAPTER 10: pages 133 to 159

[1] As quoted by Dame Kathleen Lonsdale in a Discourse given at The Royal Institution on 23 January 1970 entitled 'Women of Science': RI *Proceedings*, **43**, 297.

[2] As quoted by WHB; *Dundee Evening Telegraph*, 30 November 1931.

[3] See also Chapter 3, page 42 concerning the trip to England.

[4] *South Australian Register*, 9 May 1899.

[5] Reported in *The Register*, 30 June 1903.

[6] *The Advertiser*, 21 August 1903.

[7] *The Register*, 12 January 1909.

[8] *Brisbane Daily Mail*, 13 January 1909.

[9] *The Register*, 12 January 1909.

[10] *Brisbane Daily Mail*, 13 January 1909.

[11] W. T. Astbury was the 'keen young man' who went to Leeds as lecturer in textile physics in 1928 on WHB's recommendation. There his research

broadened into new fields to which he gave the name of molecular biology. See how it all started with a fibre in the DF, p. 124.

[12] Sir Harry Melville, *The Department of Scientific and Industrial Research*, London: Allen & Unwin 1962, 80.

[13] Address by Special Visitor Sir William Bragg to the Assembly of Faculties of University College, London 4 July 1935. *Annual Report, University of London, University College, 1935–36*, p. 113.

[14] William Bragg, 'History in the Archives of the Royal Society'; Pilgrim Trust Lecture to the National Academy of Sciences. *Science*, Washington, 89, 445–53. Lecture delivered 24 April 1939.

[15] *English Association Bulletin*, 75, December 1932.

[16] Sir William Bragg, 'Chemistry and the Body Politic'; 7th S. M. Gluckstein Memorial Lecture 1935. *The Institute of Chemistry of Great Britain and Ireland.*

[17] *The Royal Society Presidential Address* 1938; last paragraph.

[18] *Sheffield Telegraph*, 6 September 1931.

[19] *The Daily Telegraph*, 3 May 1934.

[20] It seems fair to compare this venture to the lectures to civil servants given at the RI at the Government's request in the 1960s. WL had the task of devising rather elementary science lectures which could be tactfully presented to established civil servants.

[21] See S. W. Roskill (1974). *Hankey: Man of Secrets* 3, 320. London: Collins.

[22] WHB had already expounded this idea in his *Presidential Address to the British Association* in 1928, illustrating it from his son's experience in getting sound-ranging to work in practice during World War I.

> What precisely was the difficulty which could only be resolved by a combination of knowledge and of being on the spot? It was really the difficulty of making a true estimation of quantities... It is so easy to talk generalities...and so difficult to get down to the details which make the effort a success. It may be the last little adjustment of magnitude that turns the scale.

[23] WHB's opinions had not really changed since 1916 when he had expressed the desire to see a minister for science; the authority and machinery of ministries had changed. Whereas in governments of the earlier years of this century ministers had had considerable independence, by 1939 they had become very restricted and subordinate. Thus a minister for science in 1939 would have been very much more subject to the winds of political differences than would one in 1916.

[24] *The Journal of Commerce*, 7 June 1940.

[25] *The Nation's Larder and the Housewife's Part therein*, London: G. Bell and Sons (1940), with a preface by Sir William Bragg.

[26] Public Record Office: Cab. 21/829, Royal Society Confidential Correspondence. The file starts with this letter and includes all the other Cabinet documents quoted in the text except where otherwise stated.

[27] The work of the Scientific Advisory Committee (SAC) started briskly with nine meetings in the first two months. WHB wrote to Hankey on 16 October 1940 saying: 'the sooner we have concrete proposals in mind — any of us – the more quickly we shall get hold of our task... Our main

purpose is of course to gather together the scientific ability and knowledge that can be of service to the nation and to ensure its being used, so far as it is in our power to do so.'

WHB suggested getting key scientists to come and talk to the committee, and Hankey wrote round to ministers asking for reports on scientific activities being carried on in their Departments. One of the most useful aspects of the SAC became the passing on of information about experimental work being done by different groups, so avoiding duplication and waste of time. Consultation with American scientists (before America was in the war) and the pooling of results with them was a valuable extension of the same idea.

WHB in the same letter to Hankey (quoted above) recommended thorough investigation into scientific manpower and the use of alien scientists in this country; and, always with an eye on education, wrote 'It is within our province to consider in a broad way what science is taught and how it is taught. . .' because so many young scientists would be wanted. Looking to the future he maintained 'We have too to bear in mind that we are taking [a] rather longer view than the executive departments who can hardly think of anything but the war. We do not want the nation to be let down with a bump when its war activities cease; we have to think of the transition to peace times, and our body is extremely well suited for the consideration of how science can ease the change.' (Public Record Office: Cab. 90/1).

WHB's *ex-officio* membership of the SAC ended when he was succeeded by Sir Henry Dale as President of the Royal Society at the end of 1940; however he was asked to stay on the committee, indeed his membership was twice extended and he was still a committee member when he died in March 1942.

In the same month Lord Hankey was succeeded as chairman by R. A. Butler. By that time the committee had accomplished much good work under Hankey's able chairmanship, but had not counted in the counsels of State as much, perhaps, as it might have done. Stephen Roskill in his biography of Hankey puts this down largely to the influence of Lord Cherwell, who had opposed the formation of the SAC and preferred to influence Churchill personally rather than have him take advice from this new committee. For instance, although the SAC had advised furthering the development of nuclear fission in October 1940, it was not told that the 'Maud' Committee to investigate the feasibility of producing an atomic bomb had already been set up in the June before, and their findings were not officially communicated to the SAC until July 1941. The SAC then set up a special Defence Services Panel to consider the 'Maud' recommendations, but though this Panel produced its 'authoritative and comprehensive report' as early as September 1941, the decision had already been taken to proceed with the development of the atomic bomb. Churchill preferred swifter and more secret ways of getting the job done than by committee work.

Many felt at the time that the SAC should have been used more. One of Butler's first acts as chairman was to prepare a letter to send out to colleagues in the government 'asking for your help in making the work of the Committee as effective as possible'. He refers to the Prime Minister's minute at the inception of the SAC in 1940 and continues 'You will see, therefore, that in a large part of our work, we are dependent upon the initiative of the departments in bringing problems before us. I know, of course, that you have your own channels through which you obtain scientific advice on many subjects, but there may be questions on which the assistance of a small Committee, such as mine – combining representatives of the Royal Society and of the three Research Councils – would prove useful without any overlapping with other bodies.' This seems almost like touting for work. However, in a letter to Sir John Anderson written on 17 July 1942, Butler claims that the SAC had 'acted as a lightning conductor for the recent anxiety on the scientific conduct of the war'.

The SAC was always liable to meet obstruction from individual departments who relied upon the advice of their own attached scientists. After Butler resigned the chairmanship at the end of 1942 the weight of this central committee declined rapidly, and it did not survive to influence the peace as WHB and his colleagues had hoped it would.

For a fuller account of the formation and activities of the SAC see Stephen Roskill's biography *Hankey: Man of Secrets*, 3, London: Collins 1974.

CHAPTER 11: pages 160 to 170

1 *The Washington Post*, 25 April 1939.

2 See note 14 to Chapter 10.

3 Thomas à Kempis, *The Imitation of Christ*, translation by L. Sherley-Price, 1952, p. 31. Penguin Classics.

4 Sir William Bragg (1920), *The World of Sound* 195, London: G. Bell & Sons.

5 The analogy of having teeth extracted without anaesthetics was one which must readily have come to WHB's mind: his Uncle William (the chemist) operated thus on his shop counter.

6 Twenty-five years before this, the philosopher F. D. Maurice had declared that it was contrary to the revelation in the Bible that 'eternal' should mean time everlasting. 'Eternal life' meant the present reality of knowing God and Jesus Christ; separation from them being 'eternal death'. The thought of theologians is slow to permeate parochial congregations and, anyway, Maurice was dismissed from King's College in 1853 for such views. Belief in everlasting punishment was a pillar of Victorian orthodoxy. (See A. R. Vidler, 1961, p. 86: *The Church in an Age of Revolution:* Pelican History of the Church 5).

7 Original manuscript by WHB in the author's possession.

8 Quotation from Stewart Headlam by A. R. Vidler, 1961: op. cit. p. 119.

9 Report in *The Morning Post*, 31 January 1924, on address delivered by WHB at Sir John Cass Technical Institute. For longer excerpt see p. 130.

10 *America*, 29 September 1928.

[11] *Penang Gazette*, 5 April 1934, quoting from *The Daily Express*.
[12] *The Morning Post*, 8 June 1936; Series 'What Next?' No. 1. 'Science and Mankind' by WHB.
[13] *The Star*, 29 March 1939.
[14] *Mining and Metallurgy*, 7 February 1942.
[15] *The Hibbert Journal* xxxviii, 3 April 1940.
[16] Compare A. R. Vidler, 1961, op. cit. p. 24.
[17] Sir William Bragg, *Science and Faith:* University of Durham, 13th Riddell Memorial Lecture 1941, London: Humphrey Milford.

CHAPTER 12: pages 171 to 178

[1] Sir Lawrence Bragg, 'Science and the Adventure of Living'; 4th Radford Mather Lecture delivered at The Royal Institution 25 October 1950. *The Advancement of Science,* **VII**, No. 27.
[2] *The Spectator* 3 October 1928.
[3] W. H. Bragg, 1942 'Science and the Community', *Endeavour* **I**, No. 1.

APPENDIX: LETTERS AND BROADCAST

A. Letter dated 5 February 1962 from W. L. Bragg to Professor Paul P. Ewald FRS New Milford. Conn. USA (see p. 74)

The Royal Institution
21 Albemarle Street W1
5 February 1962

Dear Ewald,

I have your letter and notes about my father. . .

You ask how he started on X ray work. I am practically certain it was because Laue sent him a copy of his paper, knowing his interest in X rays. My father had this paper when we were staying with friends in Yorkshire in July or August 1912, and he and I discussed it often. I of course warmly upheld his theory of corpuscular X rays. My recollection is that I suggested the 'avenue' theory to him. When we got back to Leeds I set up an experiment to test it. I passed a beam of X rays through a collimator, and then on a small crystal which closed the aperture of a light-tight box. At the back of this box was a photographic plate. The idea was to vary the orientation of the box throughout a solid cone, and see if the X rays shot down the avenues and made spots when an avenue became parallel to the beam. Since crystal and plate moved together, each avenue would register in a constant position on the plate.

I got no results. If only I had had a strong X ray tube with a copper target I would have got something very like a rotation photograph! But with a feeble platinum cathode tube there was just a mess. . .

As to the steps by which he [WHB] became convinced that X rays were being diffracted like waves, I think they were these:

He was not convinced by Laue's paper; I remember he had a large wooden globe on which he plotted Laue's spots and their geometrical relationships, and was trying to think of ways in which they could be explained by the neutral pair theory.

On my return to Cambridge I restudied Laue's results and explained the ZnS spots as being due to diffraction by a face-centred cubic lattice (Oct. 1912). This, I think, convinced my father that they were diffraction effects, but he still thought that there might be electro magnetic waves associated with the X rays and in deference to his views I called my paper (Nov. 1912) 'The

Diffraction of short electro magnetic waves by a crystal'. My father was still wavering when he wrote the note to *Nature* on Nov. 28th which you quote. But just about that time I showed that X rays were specularly reflected over a range of angles by the cleavage planes of mica (*Nature*, Dec. 12th 1912) and my father at once examined the rays reflected from mica with an ionisation chamber and convinced himself that they were really X rays. This led to his designing the X ray spectrometer. I had no hand in planning the spectrometer, but while this was going on I had worked out the structure of NaCl and KCl by Laue photographs, and that is how my father came to use rock-salt for the curves in our first joint paper 'The Reflexion of X rays by Crystals', on which we worked together during the Xmas holidays of 1912–13.

These results of course showed the immensely greater power of the spectrometer for analysing crystals. I was very much teased at the time for upsetting my father's theory! But you will see how much my father's work at Leeds, and mine at Cambridge, were interwoven at that early time. . .

With kind regards,

Yours sincerely

W. L. Bragg

B. Letter dated 13 September 1927 from W. H. Bragg to Sir Oliver Lodge (see p. 149)

My dear Lodge,

Your letter is extremely interesting. I think it is quite a privilege to hear at first hand of the way in which the meetings were conducted in the old days. I doubt and I have many regrets in doing so whether it is possible to recapture the old flavour. Things have changed and I have often thought that in planning future proceedings at the BA we must go back to the first principles. Discussions such as you describe do not arise naturally in these later days. You yourself mention some of the reasons – extreme specialisation for one.

There is another reason which seems to me of importance: the BA was the only place where such discussions were carried on. It is not so now. There are many bodies eager to promote the advance of science by discussions. We might rightly say I think that the BA has been the pioneer and exemplar. If, for example, the subject of colloids is to be discussed by experts it is done at a special meeting at, let us say, the Faraday Society. I doubt if it is ever likely that the experts will meet again at the BA for the purpose of discussing a subject among themselves. Discussions take place but they are arranged for and are in a sense artificial. The subject of colloids was not really discussed at all at the last meeting. The whole affair was intended I believe to give instruction to the junior men and especially to those who are interested in applications to textiles.

It seems to me that you might put discussions into four grades:

No. 1. The highest. A real argument between those who are expert or at least have fresh ideas to communicate. The object being to clarify each others' thoughts and weld something new out of the accumulated materials.

2. A setting forth of a subject, hardly to be called a discussion, which is intended to disseminate expert knowledge among workers in other subjects as well as that to which the subject really belongs.

3. Discussions intended for the benefit of the general public.

4. Popular lectures intended to interest people in scientific subjects.

I am not sure whether No. 1 was included among the original purposes of the Association? It was a happy thing no doubt but it became a practice. That is clear from your letter. When there are so many bodies now-a-days that devote themselves to purposes like No. 1 the BA's usefulness in this respect seems to have mainly disappeared: the work can be better done elsewhere. Moreover a good many think as I do that there are already quite enough of these conferences.

The second point is really something to be considered. It rather corresponds to the purposes of the Royal Institution but the audience can be much larger and wider though the meetings are less frequent. I may say that even in this respect times have changed. The Royal Institution was once unique in giving such lectures in London. The various University bodies now pour out a flood of lectures and one has to think carefully what it is that the RI can do and that the Universities do not do in the way of lecturing.

The third point, I suppose, is really the main feature of the BA or at least was what was first intended. I rather think that it is still the most important purpose that the BA can fulfil. 'The Advancement of Science' includes the advancement of the appreciation of science. For a short period every year the BA has the ear of the whole public and it is then that the opportunity comes.

As to the last point, I need not say much. Popular lectures are very useful and the BA does its best to give good ones.

I have written this letter mainly as a contribution to a discussion. I do not assert the opinions in it as final on my part but I am very interested in the endeavour to think out what we are really aiming at in these times at the BA meeting.

Yours sincerely,
W. H. Bragg.

C. Broadcast to Germany in January 1939: Help to Save Civilisation (*see p. 115*)
'*This statement, signed by 18 outstanding Britons and made public last night, was broadcast by the BBC in German*' (*The Manchester Despatch* 28 *January* 1939)

A spirit of uneasiness broods over the world. Men and women in every country are uncertain what the next weeks and months may bring.

They see huge armaments piling up on every side, they see plans being made for civilian defence, and they realise only too vividly that war under modern conditions between highly organised States can bring no good, but only death and destruction to countless homes, irrespective of age or sex. They see our civilisation, to which men and women of all classes and in all countries have contributed, threatened with the greatest catastrophe in human history.

It is time, if we are not to be too late, that men of good will who value the fruits of civilisation, who have no hatred or spirit of revenge in their hearts, and who desire in all sincerity to live on terms of friendship with their fellow men in every country, should speak across the frontiers to those who feel as they do in order that they may use together their gifts of heart and mind to co-operate in preventing the supreme catastrophe and in breaking down the artificial barriers of hatred by which we are in danger of being divided.

We in Britain have no desire to dictate to others. While resolutely determined to maintain our own liberty we stand for peace, a peace of equality for all and of justice for all.

We stand for the rule of law in the relations between States, the only basis on which our civilisation can be preserved. We recognise that no civilisation, if it is to survive, can be static, but no nation will find a lasting solution for its problems save in a spirit of co-operation with others.

We appeal above all to leaders and people in the great German Reich at this moment of power and influence in their history.

We appeal to them to use those gifts by which they have for centuries enriched our common heritage in all fields of human knowledge and activity, and to join with us in a supreme effort to lay the spectre of war and enmity between nations, and in a spirit of free and willing co-operation, by which alone can their needs and ours be satisfied, to build with us a better future so that we may not only preserve civilisation, but hand it down to our children enhanced by our experience.

The signatories are:

Marquis of Willingdon
Earl of Derby
Lord Dawson of Penn
Lord Horder
Lord Macmillan
Lord Stamp
Mr. Montagu Norman
Mr. H. A. L. Fisher OM
Mr. G. M. Trevelyan OM
Lord Eustace Percy
Sir Michael Sadler
Dr. Vaughan Williams OM
Sir William Bragg OM
Sir Arthur Eddington
Sir Edwin Lutyens PRA
Sir Kenneth Clark
Mr. John Masefield OM
Lord Burghley

D. Letter dated 10 June 1940 from W. H. Bragg to The Prime Minister, the Rt Hon Winston Churchill CH (*see p. 152*)

The Royal Society
Burlington House,
London W1
10 June 1940

Dear Prime Minister,

The appointment by the Lord Privy Seal of the Committee to advise on Food Requirements and Production, of which I have been invited to act as Chairman, affords clear evidence of the Government's determination to utilise all the resources of science in solving the problems of the war. If these resources are to be applied quickly and effectively, not to this problem alone, but to all those which must be faced if we are to achieve victory, it is my own view, and that of many of my colleagues, that there is urgent need of a piece of co-ordinating machinery which does not at present exist. In the Department of Scientific and Industrial Research, the Medical Research Council and the Agricultural Research Council, the Government has at its disposal three official bodies, each fully competent to give advice within its own sphere of activity, and covering between them some of the most important fields of science in which the national activity is engaged. There are, however, many problems that concern more than one of these bodies – food requirements and production provide a case in point – and to be fully effective they need to be in constant touch, not only with one another, but with the various Government Departments, and with the Cabinet. The need for such co-ordination, so far as scientific research is concerned, was fully recognized by the Research Co-ordination Sub-Committee which reported in 1928; but the machinery then set up has since largely lapsed, and a method which may have been adequate to the needs of peace-time is quite inadequate in time of war. It is not to-day a question of planning long-term research, but of the immediate utilisation of existing scientific knowledge, and of the rapid solution of practical scientific problems. It seems that a focus point of the kind proposed in this letter is at present lacking in the organisation of our national scientific effort.

If it is to work quickly and effectively, any committee entrusted by the Government with these co-ordinating functions must be small, and must have direct access to the Cabinet. It should, if possible, have a Cabinet Minister as its Chairman, and a member of the Cabinet Secretariat as its Secretary. I would suggest, for your consideration that an effective co-ordinating committee might be constituted as follows:

Chairman.
Vice-Chairman. The President of the Royal Society.
The Physical Secretary of the Royal Society.
The Biological Secretary of the Royal Society.
The Secretary of the Department of Scientific and Industrial Research.
The Secretary of the Medical Research Council.
The Secretary of the Agricultural Research Council.
Secretary.

A committee so constituted would have the following advantages:

(1) Its chairman, as a member of the Cabinet, would be able to consult the committee on problems which require scientific advice for their solution and on which the Cabinet as a whole, or any Department of the Government, required such advice. He would also be able to convey to the Cabinet any suggestion of the Committee as to the possible use in war of scientific knowledge or methods not yet fully exploited.

(2) The presence on the committee of the Secretaries of the

Department of Scientific and Industrial Research, the Medical Research Council and the Agricultural Research Council would ensure that the wide resources possessed by these bodies were brought to bear on any problem affecting more than one of them, and that there was no overlapping or wastage of effort.

(3) The presence on the committee of the President and the two Secretaries of the Royal Society would help to focus on our common war effort all scientific activities, including those not now represented in any Government Department.

(4) Such a committee would be in a position to advise the Cabinet as to those scientists best fitted to apply existing knowledge to any war problem. By co-ordinating scientific work they could ensure economy in the use of both workers and equipment.

(5) If the Directors of Scientific Research of the Defence Services, and members of the scientific staff of any other Government Department, acted as assessors to such a committee when problems affecting their departments were under discussion, it would be possible to ensure that all such problems were attacked, without loss of time, by those best fitted to solve them.

The duties of such a committee should be:

(1) To make recommendations on scientific matters referred to them by the Cabinet, and in particular to advise as to the existing bodies, or new *ad hoc* committees, that could most quickly and effectively solve any such problem under consideration by the Cabinet.

(2) To transmit to the Cabinet any suggestions as to the better utilisation of existing scientific knowledge in the prosecution of the war, or as to any scientific problems that demand urgent study because of war conditions.

(3) To keep the Cabinet constantly informed as to the progress of any action or enquiry arising from (1) or (2), and as to the steps that should be taken to expedite such action or enquiry if progress is regarded as unsatisfactory.

(4) In all these activities the functions of the committee should be limited to effective co-ordination. It should not in any way usurp the existing functions or facilities of the Department of Scientific and Industrial Research, the Medical Research Council, the Agricultural Research Council, or any other official bodies. It should be concerned solely with rendering rapidly effective the activities of these bodies, and of any other official or unofficial scientific workers or institutions, by eliminating waste of effort, isolated action, or any other obstacle to rapid progress.

You will, I am sure, believe that I am making these suggestions only with a view to ensuring that science is allowed to make its most effective contribution to the prosecution of the war.

Yours sincerely,
W. H. Bragg
President, Royal Society.

The Rt. Hon. Winston Churchill, CH

E. Letter dated 24 July 1940 from W. H. Bragg to Sir Alan Barlow CB The Treasury SW1 (see p. 153)

The Royal Institution,
21 Albemarle Street,
London W1
24 July 1940

Dear Barlow,

At the interview which you kindly gave to Egerton, Topley and myself a fortnight ago, you asked for examples of cases in which the committee proposed by us would have been of use to the Government. I have been thinking much about your enquiry.

If we try to give details of particular instances we come up against the difficulty that to be effective in this way we must accuse individuals of neglect and inefficiency. It would be a horrible task to get up a case involving the preparation of such evidence.

Yet your question was a very proper one; and if we cannot answer it in one way we must try another. I think that this can be done by keeping to more general questions, strategy rather than tactics. Here, for example, is such a general matter: –

By the end of the last war the main body of scientific men had been mobilised, and their special knowledge and inventiveness were being used in the Navy, the Army and the Air Force. I do not exaggerate when I say that their work was not merely contributory to the success of the Allies but was essential to it. When the war ended the best of the men returned gladly to their civil work. The men who remained to carry on the scientific development of the various weapons and defences were generally less able, less inventive, less brilliant than those who had gone. There were notable exceptions of course. Also their work might have been better encouraged and more generously treated if they had been better scientists and more pushful men. There have been energetic advances in some directions since then, e.g. in the matter of aeroplane design and construction, and in asdics. But there have been failures also, e.g. in the neglect of the technique of explosive maufacture, or again in sound ranging where instruments have been 'improved' till they have become impracticable. Take the former of these two. After 1914 Mr K. B. Quinan, CH, from his experience in erecting and operating the enormous factories for explosives saw that there was a grievous lack of chemists with chemical engineering experience. On account of his representations, action was taken which resulted in the erection of the Department of Chemical Engineering at University College, London, in addition to that existing at Imperial College. This was a step forwards, but it has not been so effective as had been hoped. During the years after the war certain Ordnance Factories were kept in being at small production. But they do not seem to have taken advantage of the Chemical Engineers that were being trained at the two Colleges mentioned, nor was encouragement given to the engineers who are so very necessary to the explosives industry.

It may have been no one's business to call attention to defects such as this, but it is certain that the body we suggest would have got to know about it in some way, and would have called attention to a situation which was dangerous and has in fact turned out to be extremely serious. Chemical knowledge and its applications to explosives technique have greatly advanced since 1918; yet it is only now that the consequences are being worked out because recommendations based on that knowledge have not been followed up. I am treading on difficult ground here, and hope you will not let me get into trouble unless it is necessary. It is not only in matters of warfare that scientific knowledge is overlooked and scientific recommendations are put aside. It was a good move when the Research Councils for Medicine, Agriculture and Industry (DSIR) were formed. But there are complaints, even in these limited fields, that suggestions are frequently neglected.

The proposed committee would help sometimes to quick decisions. Perhaps I may quote my experience on the Scientific Food Sub-Committee: –

The Scientific Sub-Committee on Food Policy provides an example of the value of a strong committee of scientific experts in helping towards decisions on matters where scientific knowledge is a determining factor. It has presented a short report on 'Bread' which has at once been adopted by the Minister of Food; the result being that important action can be taken before delay makes it useless. The report will certainly not be approved by everyone because views on nutrition run to extremes in several directions. The point is that a reasonable policy has been decided on – which *may* be wrong, but it has some of the best backing that can be got – and indecision comes to an end. Talk is hushed and work is done. The strength of the sub-committee has made this possible; it can be quoted by the Minister as authoritative.

The Committee we propose would, on a far wider scale, *provide* authoritative opinion from the great body of scientists whose work the Committee knows well. On occasion it might be able to express opinion without going further, but its chief work would be to collect such a verdict that action could be taken *promptly*. It is continually pointed out that democracies suffer from long arguments and indecisions arising from the fact that it is extremely difficult to comprehend all the factors of questions that require solution. The authoritarian state tends to decision without enquiry; the democracy tends to enquiry without decision.

After this attempt at illustration I may restate briefly the main objects of the proposed committee. It could be used:

(1) To *keep a watch* on the way in which the Nation's scientific and technical organisation copes with the problems of the day and to advise on what might be done towards improvement of its efficiency.

There is at present no body responsible to keep a watch on the organisation *as a whole*; the DSIR does its job, the Royal Society settles its affairs, the MRC looks after medical research, but there is no means by which their efforts and those of other bodies can be concerted towards the solution of specific and urgent problems. It is only by chance from a feeling of a common need or through individuals who happen to be on more than one Council that the several bodies act in harmony. There is no machinery available and the Committee could be that machinery.

(2) To *convey* to the Lord President information as to any particular scientific or technical development that the responsible scientific authorities consider need special attention.

(3) To *advise* when required as to the right body or the right person to deal with scientific or technical problems and to avoid overlapping.

[4] A body such as we wish to be created (or rather resuscitated) has not authority by reason of the individuals forming it, but by reason of numerous individuals and activities which are *represented* by the responsible officials composing it. We venture to think no ad hoc committee, much less a single individual, could

successfully carry out its functions. The Society's prestige is used by Government when it needs its help but no power is afforded to it to exercise the assistance it could and would gladly give.

(5) To *advise* also on the setting up of scientific and technical committees, or the appointment of members of such committees or of individuals who may be required to deal with matters of a scientific and technical kind.

There are instances of appointments being made without any guidance from authoritative scientific bodies and chosen without due regard to scientific opinion. A committee such as that proposed could obviate the difficulties which are being caused by such appointments.

Yours sincerely,
W. H. Bragg

Sir Alan Barlow, CB,
The Treasury,
Whitehall, SW1

INDEX WITH BIOGRAPHICAL NOTES